THE WAY IT WAS

MATHEMATICS FROM THE
EARLY YEARS OF THE
BULLETIN

DONALD G. SAARI
EDITOR

AMERICAN MATHEMATICAL SOCIETY
PROVIDENCE, RHODE ISLAND USA

2000 *Mathematics Subject Classification.* Primary 00B60.

For additional information and updates on this book, visit
www.ams.org/bookpages/bulhig

Library of Congress Cataloging-in-Publication Data
Saari, Donald G.
    The Way It Was: Mathematics from the Early Years of the Bulletin / Donald G. Saari, editor.
        p. cm.
    Includes bibliographical references.
    ISBN 0-8218-2672-7 (alk. paper)
    1. Mathematics. I. Saari, D. (Donald) II. Bulletin (new series) of the American Mathematical Society.

QA7.M324224   2003
510–dc22   2003063303

Tatjana,

the newest star, and born in the far North!

Contents

CONTENTS

Preface

Perhaps it was serving as the Chief Editor of the *Bulletin* (1999-2005) that generated my interest in the early years of this journal; perhaps it was as a mathematician wanting to learn more about the history of American mathematics. From wherever the itch, old issues of the *Bulletin* provided a satisfying scratch. Selective articles from the *Bulletin* and its predecessor the *Bulletin of the New York Mathematical Society* are reproduced here.

I am a mathematician, not a historian, so, like a novice stumbling through an art gallery "knowing what I like," these articles reflect what interested me. While selected to offer a flavor of the time, many of them have historical significance. So, in each chapter, my introductory comments suggest the importance of the articles while describing a bit about the mathematicians, particularly those who are not house-hold names.

My interest in the history of our field got out of hand with a bad habit developed over my many years at Northwestern University. Before the time when "senior books and journals" were relegated to the "old-books home" of cold storage, those older journals, craving attention, seemed to jump out and grab you when walking by. So, when going to our math library for a specific purpose, it was so easy to succumb to the overwhelming temptation to spend *just a few minutes; well, maybe a few more minutes,* rummaging through old math journals. The seed was sowed at Northwestern, but this project was started after moving to the University of California, Irvine. A problem being at a school younger than you are (UCI was founded in 1965) is that old *Bulletin* issues are not readily available. Therefore, my reading of old *Bulletins* was, and is, done during the summers in the Upper Peninsula of Michigan near Michigan Technological University. My thanks to the Head of their Mathematics Department, Al Baartmans, for arranging for library privileges. Also my thanks to Bill Sved, the Asst. Dir. of Research Services in the MTU library, who cheerfully helps find these old *Bulletin* volumes. Thanks also to Sergei Gelfand from the AMS for his encouragement — and a somewhat consistent but kind nagging to complete this project — and

Christine Thivierge for finding and sending me copies of several articles.

Irvine, California
September, 2003

Chapter 1

Introduction

The January 2000 issue of the *Bulletin of the American Mathematical Society* starts with the quote

> *"The extraordinary development of mathematics in the last century is quite unparalleled in the long history of this most ancient of sciences. Not only have those branches of mathematics which were taken over from the [nineteenth] century steadily grown but entirely new ones have sprung up in almost bewildering profusion, and many of these have promptly assumed proportions of vast extent."*

This comment states the obvious; we all know and even have experienced this. With the rapid changes, stunning advances, emergence of new research areas and the tumbling of named problems over just the last twenty or thirty years, it is clear that we live in one of the most exciting periods of history to be a mathematician. What underscores the continuing vitality and growth of our discipline is that by changing the bracketed "nineteenth" to "eighteenth" we recover the opening paragraph of James Pierpont's 1904 *Bulletin* article where he reviewed the progress of *nineteenth* century mathematics.

The year 2000 provided an appropriate juncture to review where we are and how we got here. In this spirit, selected *Bulletin* articles and book reviews published during the first quarter of the twentieth century were republished in the January 2000 issue. That special issue, with articles by Birkhoff, Brouwer, Einstein, Fiske, Pierpont, Poincaré, and Van Vleck, along with the well written book reviews by so many eminent names from the history of our field, only sampled the development and early days of the *Bulletin* and American mathematics. The positive reaction received by this special issue suggested that it wetted the appetite for more. This book, with reprinted articles from the *Bulletin* and its predecessor *The Bulletin of the New York Mathematical Society* during the 1890s and first years of the 1900s, is a response. The articles selected are intended to capture the spirit of mathematics during that formative time; they sample the mathematical issues, advances, and struggles of a century ago. What emerges is a picture of the change and growth in mathematics that supports Pierpont's enthusiasm.

To experience mathematics of that period before automobiles were developed and prior to the inaugural Kitty-Hawk flight, put on old clothes — this is necessary because you will get dirty — and venture into the dusty reaches of a mathematics library to thumb through old issues of the *Bulletin* and its predecessor. Come equipped with a knife or scissors to separate tear sheets because, in several cases, you will be the first to browse many of the articles; you may even be the first to open some of the volumes. Handle them with particular care as pages may crumble and the binding of many volumes may be ready to collapse. But the time spent will be enjoyable; reading book reviews, articles, obituaries, and meeting reports will transport you to the mathematical community — its advances, and concerns — of a century ago. Let me offer a sampler of what can be found before, in the next chapter, introducing the first articles that are reproduced here.

Growth and Changes

Even skimming these old articles, the flow of concerns and issues takes new meaning when expressed in the words of the time. As an example, we may have an idea how the the movers and shakers of our field, the top experts, considered mathematics, but how did a rank-and-file mathematician view this subject a century ago? Remember, the American mathematical research community was in its infancy and far behind Europe. This status is reflected by several papers indicating that topics and approaches currently taken for granted still carried a sense of bafflement; at least in the American mathematical community. A taste of the struggles in accepting abstraction jumps out of Simon Newcomb's [7] 1893 address to the New York Mathematical Society.

To set the stage, recall the story of the person, well over a century ago, resigning his position from the US Patent office because everything that could be invented had already been invented. Mathematics had its own nay-sayers. In his talk, Newcomb observed that

> *"Among the miscellaneous reading of my youth was a history*
> *of modern Europe, which concluded with a general survey and*

*attempted forecast of progress in arts, science, and literature.
...On the subject of mathematics the writer's conclusion was
that fruitful investigation seemed at an end, and that there was
little prospect of brilliant discoveries in the future."*

So much for this unnamed person's forecasting abilities. What provides
a glimpse into the thinking of American mathematicians of the 1890s is
Newcomb's further comments. He argues that if one takes a particularly
narrow perspective,

> *"I am not sure that careful analysis would not show the author's
> view to be less rash than it may now appear. May we not say that
> in the special direction and along the special lines which math-
> ematical research was following a century ago no very brilliant
> discoveries have been made? Can we really say that Euler's field
> of work has been greatly widened since his time? Of the great
> problems which baffled the skill of the ancient geometers, includ-
> ing the quadrature of the circle, the duplication of the cube, and
> the trisection of the angle, we have not solved one. Our only
> advance in treating them has been to show that they are insolu-
> ble. To the problem of three bodies we have not added one of the
> integrals necessary to the complete solution. ...For the general
> equation of the fifth degree we have only shown that no solution
> exits. We should, doubtless, solve many of the problems which
> the Bernoullis and their contemporaries amused themselves by
> putting to each other, rather better than they did; but, after all,
> could we get any solution which was beyond their powers?"*

Newcomb is setting up his audience to point "...out in what direction
progress lies and what is the significance of mathematical investigation at
the present day." He is more encouraged about the future of American mathematics than he had been a couple of decades earlier (see Karen Parshall's
excellent survey [4]), and one just feels how he wants American mathematicians, still at an early stage of understanding research, to appreciate how
some of the power of mathematics derives from its ability to unite seem-
ingly disparate concepts. He introduces the more modern, abstract way of
thinking of that time where

> *"progress has been made by going back to elementary principles,
> and starting out to survey the whole field of mathematical in-
> vestigation from a higher plane than that on which our prede-*

cessors stood, rather than by continuing on the lines which they followed."

Newcomb's comments reflect his recognition that many American professional mathematicians of the 1890s were unsure of the new directions. Rather than as an adjective, "abstract," from "abstract mathematics," needed to be treated as a verb indicating *what* is being abstracted — which is not a bad idea. As the utility of these approaches still needed to be described and justified, Newcomb provided several examples to illustrate the unifying power of this approach creating

> *". . . a wide field before us, embracing the first conception of groups, and with it an important part of our modern mathematics. . . .*
>
> *The idea of groups of operations, as I have tried to develop it, has in recent years been so extended as to cover a large part of the fields of algebra and geometry. Among the leaders in this extension has been Sophus Lie."*

But, how can we not be amused by Newcomb's colorful words used to reflect the fascination of the time with higher dimensional geometry.

> *"Another field of mathematical thought . . . may be called the fairy-land of geometry. To make a mathematician, we must have a higher development of his special power than falls to the lot of other men. When he enters fairyland he must, to do himself justice, take wings which will carry him far above the flights, and even above the sight, of ordinary mortals. To the most imaginative of the latter, a being enclosed in a sphere, the surface of which was absolutely impenetrable, would be so securely imprisoned that not even a spirit could escape except by being so ethereal that it could pass through the substance of the sphere. But the mathematical spirit, in four-dimensional space, could step out without even touching any part of the globe. Taking his stand at a short distance from the earth, he could with his telescope scan every particle of it, from centre to surface, without any necessity that the light should pass through any part of the substance of the earth."*

A fourth dimension? What is this? Does it have any meaning? During this period prior to relativity and string theory, *"[f]rom the point of view of physical science, the question whether the actuality of a fourth dimension can be considered admissible is a very interesting one. All we can say is that, as so far as observation goes, all legitimate conclusions seem to be against it."*

Newcomb (1835-1909) was an interesting, obviously self-driven individual. "At age eighteen ..., with no money and little education, [he] made his way on foot from his native Nova Scotia to the United States."[1] To improve his education, Newcomb studied mathematics in libraries in the Washington DC area before going on to graduate from Harvard in 1858. While his research emphasized celestial mechanics, mathematical astronomy, and theoretical dynamics, he also wrote on non-euclidean geometry, economics, astronomy, and was considered a leader in American science. His research contributions led to several honors, including the gold medal from the Royal Astronomical Society, the first winner of the Bruce medal (in astronomy), and even having a crater on the moon named after him! Newcomb was a founding member and first president of the American Astronomical Society. Prior to this, he was the fourth president of the AMS.

Names from the past

What makes reading these old *Bulletin* papers so delightful is how they put life into names that, today, often serve more as place holders or identifiers, such as Lie Series or Abelian Groups, than as results associated with individuals. They should because, at that time, they were commenting on contemporaries with reports that included personal information. One learns, for instance, that many prominent figures in our field did not intend to be mathematicians; they were attracted to our field because their talents were recognized and encouraged by others. Beyond providing a lesson about the importance of mentoring younger colleagues and emerging talent, the names of some of these "others" describe professional connections that started early on. As H. Fine [1] of Princeton wrote in his description of Kronecker's contribution to the theory of equations

"Leopold Kronecker, one of the most illustrious of contemporary mathematicians, died at Berlin of the 29th of last December [1891] in his 68th year. ... He was the last of the great triumvirate — Kummer, Weierstrass, Kronecker — to be lost to the [University of Berlin]. ... Kronecker was born at Liegnitz near

---

[1]From the 1898 Bruce Medalists citation of Newcomb.

Breslau in 1823. While yet a boy at the Gymnasium of his native town his fine mathematical talents attracted the notice of his master, Kummer, whose distinguished career was then just beginning. Kummer's persuasions rescued him from the business career for which he was preparing ... He studied at Breslau, ... [of] his instructors besides Kummer he was most influenced by Dirchlet, owing in part to Dirchlet's commanding abilities, in part to the strong arithmetical bent of Kronecker himself."

Fine's comments about their ages,

"... younger than [Kummer and Weierstrass], Kronecker was overtaken by death [at the age of 68] in the midst of the work to which his life had been devoted"

underscores the fact that many mathematicians enjoyed, and enjoy, long, highly productive careers. So, with this reality, where did that silly adage that a mathematician does his or her work before the age of thirty come from? This is not a recent descent into nonsense; similar comments can be found in early *Bulletin* issues. Wilson [15], in his 1906 review of books about Helmhotz, suggests a potential source for this comment by remarking that

"Helmholtz had a very marked ability for keeping out of ruts. A predisposition for getting snugly into some rut is perhaps one reason for the opinion current in some quarters to the effect that one's work is done at the advanced age of forty."

Well, a century ago this silly expression added another decade to a career. As Wilson goes on to describe, Helmhotz's exciting professional life was anything but a rut. This is reflected by one of the biographies reviewed by Wilson which

"... is divided into sections corresponding to the different positions which Helmholtz occupied — a student, a naval surgeon, teacher in an academy of fine arts, professor of physiology at Heidelberg, of physics at Berlin, and president of the Reichsanstalt. ... It should be remembered that Helmholtz did not have a free choice as to his profession. He always said that he was a born physicist. But his father was poor; [2] being a physicist was too

---

[2]His father, a teacher in the Gymnasium of Potsdam, had five children to support. So, rather than his first choice of studying physics, which was too expensive, Helmholtz applied for a scholarship at a medical institute (Friedrich-Wilhelms Institute) in Berlin. The scholarship required Helmholtz to spend several years as a military physician; hence his work as a naval surgeon.

dear; the boy must choose medicine as his work even though both he and his father regretted it. The result was one of the world's greatest physiologists ... "

Helmholtz, a physiologist! Yes, and a great one. He has to be one of the first to seriously pioneer aspects of the mathematical behavioral sciences; a topic that today is of growing interest. A measure of the importance of his research is that, over a century later, his work on vision and the auditory system continue to be read and referenced. For readers interested in learning more, let me suggest his fascinating book *On the Sensations of Tone* [4] (where, because he even describes how to create different tones, the book is read by musicians!), his fundamental work on the theory of psychophysical measurement [5], and his work on vision [6].

Europe and America

Other *Bulletin* articles describe the rise, and decline, of mathematics in different sections of the world. As F. Franklin [2] asserts in his eulogy of James Sylvester

"From about the middle of the eighteenth century until near the middle of the nineteenth, English mathematics was in a condition of something like torpor. The second half of the eighteenth century was one of the most brilliant periods in the history of mathematics; but the magnificent achievements of Euler, Lagrange, Laplace awakened no response on the other side of the narrow seas. It seems almost incredible that the complacent conservatism of Cambridge went so far that even the notation of mathematical analysis as used on the Continent was untaught there until about 1820. ... Nothing could show more thoroughly the insular and retrograde condition of english mathematics in the early part of this [nineteenth] century. The sticking to Newton's fluxions and dots, and the barring out of Leibnitz's differentials and d's, may be set down as a consequence of the great Newton-Leibnitz controversy; but whatever the cause, so complete a separation from the great current of European thought implies stagnation deep-seated and not easily removed. ... It is almost literally correct to say that the history of mathematics for about a hundred years might be written without serious defect with English mathematics left entirely out of account."

Strong words, but the kind one learns to expects from *Bulletin* articles of that time. Franklin goes on to say that

*"That the like statement cannot be made in regard to the past
fifty years is due preeminently to the genius and labors of three
men: Hamilton, Cayley and Sylvester."*

While mathematics may have waxed and waned in other countries, American mathematicians were still trying to create a viable research community. A serious problem, as early *Bulletin* articles remind us, was to break the isolation suffered by American mathematicians. One approach was through expository *Bulletin* articles that discussed even mathematics of a half century earlier. This was necessary because, as E. McClintock [3] explains,

> *"The celebrated tracts of Lobatschewsky and Bolyai, in which
> those writers showed what geometry might become if the parallel-
> axiom were left out, were long since translated into the chief
> languages of the continent, but have until the last year [1891]
> remained inaccessible to those whose only tongue is the English.
> The thanks of this large class are due to Professor Halsted[3] for
> supplying the deficiency in good clear style."*

McClintock then goes on to describe the history and the mathematics of these developments; this is one of many old *Bulletins* articles that are enjoyable reading.

In an attempt to capture at least a flavor of mathematics around the turn to the twentieth century, this book reproduces a small selection of these old *Bulletin* articles. For convenience, they are divided into sections with a brief introductory commentary. But, there are so many other articles that could have, and maybe should have, been included. Indeed, this project was delayed, and delayed, and delayed primarily because of the temptation to read *just one more* article. My hope is that this selection will inspire others, when possible,[4] to venture into that dusty, often forgotten, section of the mathematics library. You will have an enjoyable time!

---

[3] "Geometric Researches on the Theory of Parallels," Nicolaus Lobatschewsky, Berlin, 1840. Translated by G. Halsted, Austin, University of Texas, 1891

[4] While most American mathematicians of 1900 did not have ready access to the current research of that time, until digital copies are made available, many mathematicians of this century do have have ready access to old *Bulletin* journals. The reason is clear; there are far more departments of mathematics than available volumes.

Bibliography

[1] Fine, H., Kronecker and his arithmetical theory of the algebraic equation, *Bull of New York Math Soc.* **3** (1892), 173-184.

[2] Franklin, F., James Joseph Sylvester, 1897, 299-309.

[3] McClintock, E., On the Non-Euclidian Geometry, *Bull of New York Math Soc.* **2** (1892), 21-33

[4] Helmholtz, H., *On the Sensations of Tone*, (English translation) Dover Publications, 1954

[5] Helmholtz, H., Zahlen und Messen erkenntnis-theoretisch betrachtet, *Philosophische Aufsutze Eduard Zeller gewidmet,* Leipzig, 1887, (English translation, *Counting and Measuring*, C. L. Bryan, Van Nostrand, 1930.

[6] Helmholtz, H., *Treatise on physiological optics.* (J. P. Southall, Ed. and trans.), Dover, 1962.

[7] Newcomb, S., Modern Mathematical Thought, *Bull of New York Math Soc.* **4** (1893), 95-107

[8] Parshall, K., Perspectives on American mathematics, *BAMS* **37** (2000), 381-405.

[9] Wilson, E. Von Helmholtz, *BAMS* (1906), 112-119.

Chapter 2

Spirit of the time

Lesbegue measure. Lie Series. Fredholm Alternatives. Abelian Groups. Words that identify standard, specific terms of our trade. Thus, an unexpected and delightful bonus in reading old issues of the *Bulletin* is that it endows these names with life and personality. This should be expected; these earlier *Bulletin* articles were reporting on their contemporaries. In this first section, four *Bulletin* articles are reproduced which capture some of the flavor of this time. First, some background.

The "attainable but unattained"

Browsing through these old issues reminds us of the many contributions made during this period. A partial list includes the Borel measure, Brouwer's attack on the logical foundations of mathematics, Cantor's work on transfinite arithmetic, the work of Dehn on combinatorial topology and group representations, Einstein and special relativity, Fréchet derivatives, Fredholm's work on integral equations, Gibbs' "vectors" and the foundations of statistical mechanics, Hadamard with his work on negative curvature and his independent proof (with the other by de la Vallée-Poussin) on the prime number theorem, Hilbert's efforts to place geometry on a formal axiomatic basis and, of course, his famous list of problems, Hill and his advances in celestial mechanics provided by the still-studied "Hill equations," Klein's many contributions including his lectures in the US, the Lorentz transformation, the Lebesgue measure and integral, the work of Levi-Civita and Ricci-Curbasto developing tensors, Lyapunov's notions of stability, Mittag-Leffler's work including his star theorem in complex analysis, Peano and his curve, Pearson' s fundamental work on statistics, Poincaré and his *Analysis situs* leading to topology, his papers on fundamental groups and on celestial mechanics, Russell's paradox and his publication with Whitehead of *Principia Mathematica*, the contributions of Weierstrass including his efforts to make mathematics rigorous with ϵ, δ arguments, Zermelo and his work on set theory and his Axiom of Choice, and many, many others. The comment made by Pierpont (see the introduction and the opening paragraph of the first paper) was correct; it was an exciting time to be a mathematician.

Of the many *Bulletin* papers describing the advances of mathematics during this period and the nineteenth century, a satisfying survey is James Pierpont's [13] description of *"The History of Mathematics in the Nineteenth Century.* This paper, the first reproduced in this volume, was presented in the section of mathematics at an internationally important 1904 symposium held in St. Louis; the purpose of this symposium is described later. In reading Pierpont paper, which divides mathematics into twelve leading areas of the time, it is historically important to notice how he promotes contributions made by American mathematicians. Yes, most of the major research results were still being made by Europeans, but by identifying valued results made by American mathematicians, Pierpont calls international attention to the progress in creating a viable American research community.

Another feature which springs from the pages of these old *Bulletin* articles is the attitude our forefathers brought to their study of mathematics. For instance, while many of us privately share a sense of wonderment about the power and cohesiveness of mathematics, such sentiments usually remain topics of private conversations. During this formative period of American mathematics, however, this spirit of enthusiasm often overflowed in published writings. An example, which again reminds us of a hot mathematical topic of the time, is C. Chapman's [3] 1893 review of Sophus Lie's 1888 book with Friedrich Engel, *Theorie der Transformationsgruppen.*

> *"There is probably no other science which presents such different appearances to one who cultivates it and to one who does not, as mathematics. To this person it is ancient, venerable, and complete; a body of dry, irrefutable, unambiguous reasoning. To the mathematician, on the other hand, his science is yet in the purple bloom of vigorous youth, everywhere stretching out after the "attainable but unattained," and full of the excitement of nascent thoughts; its logic is beset with ambiguities, and its analytic process, like Bunyan's road, have a quagmire on one side and a deep ditch on the other and branch off into innumerable by-paths that end in a wilderness.*
>
> *Among the most important of the newer ideas in mathematics*

is that of the group. *In its nature it is essentially dynamic, involving the notion of operating with one thing upon another. "*

The first *Bulletin* and the American Mathematical Society

Reading these old *Bulletin* issues, one is struck by the number of articles that were written versions of expository presentations delivered at professional meetings and even university math clubs; some were merely cute observations. As an illustration, I wondered what pressing mathematical concern of the day would be addressed in the inaugural article of the *The New York Bulletin of Mathematics*. What was the theme of the paper which, arguably, set forth the American Mathematical Society?

To provide the background needed to describe this article, at the end of the eighteenth century, French mathematicians such as Jean-Charles de Borda, Marie-Jean-Antoine-Nicolas-Caritat Condorcet, and Pierre-Simon Laplace played important roles in reforming the haphazard units of weights and measurements by designing and introducing the metric system. (The meter is based on an astronomical measuring instrument designed by Borda.) Thomas Jefferson, a progressive American politician of the day, learned about this reform development while ambassador to France. But even though Jefferson embraced and promoted this system it did not take hold in the United States then or even now. Is there a reason for this resistance? The first article of *The New York Bulletin of Mathematics* by W. Woolsey Johnson[1] [8] explores this concern by describing the properties of number systems with different bases; an approach which appears to have been a novelty within American mathematical groups of the time.

> *"The comparatively small progress toward universal acceptance made by the metric system seems to be due not altogether to aversion to a change of units, but also to a sort of irrepressible conflict between the decimal and binary systems of subdivision.*
>
> *Before the introduction of decimal fractions, about 1585, no connection would be felt to exist between the established scale of numeration and the method of subdividing physical units, and it would probably never occur to any one to subdivide a unit into*

---

[1]Johnson (1841-1927) graduated from Yale and taught at a variety of schools including Kenyon College, and St. Johns College, but primarily at the US Naval Academy; he even had a commissioned rank in the Navy and retired in 1921 with the rank of Commodore. An interesting tidbit is that in 1896 one of the most influential American realist artists, Thomas Eakins, did a portrait of him.

tenths. The natural method is to bisect again and again. The mechanic prefers to divide the inch into halves, quarters, eighths, and sixteenths. The retailer of dry goods, whose unit is the yard, divides it into halves, quarters, and eighths, totally ignoring the inch. The mariner not only divides the horizontal angular space in which his course is laid down into quarters, thus recognizing the right angle as the natural unit, but divides the space between the cardinal points of the compass into halves, quarters and eights. ..."

After introducing virtues of mathematics in base eight, with side comments about binary arithmetic and other bases, Johnson concludes, most surely with tongue firmly implanted in his cheek,

"As there is no doubt that our ancestors originated the decimal system by counting on their fingers, we must, in view of the merits of the octonary system, feel profound regret that they should have perversely counted their thumbs, although nature had differentiated them from the fingers sufficiently, she might have thought, to save the race from this error."

The expository nature of this inaugural *Bulletin* article reflects an explicit decision by the society. In his presidential address delivered before the American Mathematical Society at its annual meeting, December 28, 1894, Emory McClintock [11] (the second AMS president, 1891-94) felt a need to outline and discuss the "early" history of the AMS. The Society, after all, already had reached the ripe and mature age of six.

"The New York Mathematical Society, originating in 1888, was at first not much more than a small mathematical club meeting periodically at Columbia College. The first meeting was called by a circular signed by three young men.[2] The number of those who could be expected to attend these meetings was not great, but all who were able and who were sufficiently interested to do so were invited to join the Society. It was fortunate in securing for its first president Professor Van Amringe, distinguished alike

---

[2]Thomas Fiske, Harold Jacoby, and Edward Stabler. Fiske, the founder of the AMS and its first Secretary and Treasurer, got the idea of forming the club by attending meetings of the London Mathematical Society. He was in England because his professor, J. Van Amringe, encouraged him to visit Cambridge University. As mentioned next in McClintock's quote, Van Amringe became the first AMS president. For more about the early days of the AMS, see Parshall and Rowe [4].

by scientific attainments, official eminence, and administrative ability. The professor of astronomy at Columbia was also active in it from the first. The meetings of the young Society were ... attended with more than interest, I might say with zeal. ...

The Society was ... distinguished from all other American mathematical clubs or associations by two circumstances: it was formed in and took the name of the largest city of America, and it was distinctly understood to be unconnected with any institution of learning. Suggestions came to be made that its usefulness could be decidedly increased by the publication of a periodical journal. Consideration of these suggestions by the Council led to the establishment of the Society's BULLETIN ... It was decided to be inexpedient to publish original investigations, that field being already occupied by successful American periodicals."

McClintock goes on to describe how part of the dues increase needed to support the *Bulletin* was directed toward inviting well-known mathematicians in all parts of the country to become members.

"That this movement toward enlargement [of the Society] was judicious and timely was proved at once by the rapid growth of the membership, which since the middle of 1891 has included a large proportion of the prominent mathematicians of the United States and Canada. As the Society thus became in reality an association of American[3] mathematicians as a body, the change of name effected this year [1894] was only a natural sequence. Finally, the result of the change of name has been that a number of persons, including several of the highest repute, who had not previously joined the New York Mathematical Society, regarding it as a local organization, have connected themselves with the American Mathematical Society; and I need hardly say that, if any one of prominence still holds aloof, from inattention or otherwise, his entrance at any time as a member will be greeted with a hearty welcome."

Starting from the earliest years, the goals of the Society were specified in the original Constitution, which, McClintock states

---

[3]Reflecting this reality of history and today, "America" and "American" is consistently used throughout the commentary instead of the more restrictive "United States."

> *"...reminds me that there is but one object (of the Society): to encourage and maintain an active interest in the mathematical science. ...In order to encourage and maintain an interest in mathematical science, we may say, then, (1) that mathematicians must be brought to know more about each other and concerning each other's work; (2) that their number must be increased by the encouragement of the study of the higher mathematics among the young; (3) that information should be disseminated fully and speedily concerning mathematical publications abroad as well as at home; (4) that, as regards the more important of such publications, competent critics should be induced to write and publish papers descriptive of their contents and indicating their merits or defects; (5) and that every member of the Society should be stimulated to the most successful effort possible in his own line of mathematical labor, whatever it may be."*

The fourth stated objective, more generally interpreted as communicating important advances in mathematics, remains a central objective of the *Bulletin*. A consequence is that starting already in the early years of the AMS, the *Bulletin* is an excellent resource to trace changes in mathematics both internationally and in America.

Mathematical rigor.

Main themes within the American mathematical community during the time approaching the turn from the nineteenth to the twentieth century ranged from groups, the much discussed advances on the solutions of fifth degree algebraic equations, the introduction of mathematical rigor (where there are statements such as "the notion of rigor is relative and depends on what we are willing to assume either tacitly or implicitly"), to the influence of the dominating mathematical forces of the time such as David Hilbert, Felix Klein, Henri Poincaré. All of this is found in the *Bulletin*.

For me, the lack of rigor of the time was dramatically brought forth with a F. Cajori's [2] comment that a prominent mathematician

"...did not hesitate to write

$$\cdots + \frac{1}{n^2} + \frac{1}{n} + 1 + n + n^2 + \ldots = 0,$$

simply because

$$n + n^2 + \ldots = \frac{n}{1-n}; \ 1 + \frac{1}{n} + \frac{1}{n^2} + \cdots = \frac{n}{n-1}."$$

Discouragement would hit any of us if even a particularly poor calculus student would suggest that a sum of positive values equals zero! The name of the prominent mathematician? Euler. Cajori goes on to state, "The facts are that Euler reached some very pretty results in infinite series, now well known, and also some very absurd results, now quite forgotten."

Cajori's used the Euler example to introduce the history of the "Evolution of criteria of convergence;" this is the second paper reproduced in this section.[4] As Cajori reports, these notions of convergence "...caused a considerable stir. We are told that after a scientific meeting in which Cauchy had presented his first researches on series, Laplace hastened home and remained there in seclusion until he had examined the series in his *Mécanique céleste*. Luckily, every one was found to be convergent!"

For background, Swiss born Florian Cajori (1859-1930) was an eminent historian of mathematics who moved to the United States at age 16. After spending a couple of years at the State Normal School in Whitewater, Wisconsin, he received a Masters degree from the University of Wisconsin in 1886 and his Ph. D. from Tulane in 1894. He then moved to Colorado College in Colorado Springs where he held several positions in physics and mathematics; he even served as Dean of Engineering. In recognition of his work on the history of mathematics, Cajori was appointed to a newly created chair for the history of mathematics. A reflection on the stature of Cajori and his work is that this chair, one of the first for the history of mathematics, was specially created for him by the University of California at Berkeley; another reflection is the moon's "Crater Cajori." He was president of the MAA in 1917-18 and vice-president of the American Association for the Advancement of Science in 1923.

World Travel; World Congresses

During the late nineteenth and early twentieth century, Americans traveled to Europe to learn mathematics; Europeans traveled to America to teach us. *Bulletin* articles can be found describing topics of travel ranging from Klein's important "Evanston Lectures" held at Northwestern University from August 28 to September 9, 1893 (see the 1894 volumes of the *Bulletin* and Parshall and Rowe [4] for a discussion and description of these talks), all the way to what American students should expect when traveling abroad. Collidge's [4] advice for such students is amusing when read today.

---

[4]For the reader interested in learning more about the development of mathematical rigor, I recommend Pierpont's Bulletin article [14] "Mathematical rigor, past and present" which often is cited when this topic is discussed the history of mathematics

"American students of mathematics, who intend to devote their lives to the subject, have generally in the past deemed it wise to spend more or less time at a European university, ... [and] this migration will continue. ... True it is that an American who studies at Oxford must wear academic dress at the proper times and seasons, and must forego the dangerous habit of carrying weapons such as javelins, while at Cambridge, he will not be allowed to play marbles on the steps of the Senate House ... In Italy, however, the statutes are planned for one type of student wishing one thing, and when the student presents himself who wants something else the process of adjustment is difficult and jarring."

Travel to neighboring cities involved trains, or vehicles depending upon the original version of "horse power," while a trip to Europe required a long journey across the sea. Thus it is understandable why attendance at distant meetings was rare. Faced with the reality that so few from American could attend foreign meetings, an interesting service supplied by the *Bulletin* was a sharing of the events through surprisingly complete, written reports; reports that offer us an excellent view of mathematics of that time. An example reproduced here, which provides a flavor of a particularly important gathering over a century ago, is the report by Charlotte Angas Scott [15], from Bryn Mawr College, of the 1900 Paris Congress of Mathematicians. She was one of six from the United States attending this ICM.

Beyond descriptions of talks, Scott tells us that contrary to the expectations of having 1,000 mathematicians participating, "...the total attendance can hardly have exceeded 250 in all." Her report provides an excellent sense of the meetings, the content of the talks, and she even critiques the style of presentations.

"One thing very forcibly impressed on the listener is that the presentation of papers is unusually shockingly bad. Presumably the reader desires to be heard and understood; to compass these ends, instead of speaking to the audience, he reads his paper to himself in a monotone that is sometimes hurried, sometimes hesitating, and frequently bored. He does not even take pains to pronounce his own language clearly, but slurs or exaggerates its characteristics, so that he is often both tedious and incomprehensible."

Not all presentations were full of faults; "the Italians, with their clear and spirited enunciation, come nearest to being free from them." For individuals,

Mittag-Leffler (who, at this meeting, described his star theorem for complex variables) was singled out to receive praise. Beyond compliments, Scott used him to promote an argument that is as valid and appropriate today as a century ago.

> *"It would be invidious and impertinent to mention names [of particularly bad speakers]; the special sinners sit in both high and low places. But it is perhaps pardonable to refer to M. Mittag-Leffler's presentation of his paper ... as showing in how admirable and engaging a style the thing can be done. It is not given to everyone to do it with this charm; but there is no excuse for any normally constituted human being, sufficiently versed in mathematics, failing to interest a suitable audience for a reasonable time in that which interests himself, ... "*

The historic importance of 1900 Paris Congress of Mathematicians derives, in part, because this is where Hilbert introduced his problems. (Hilbert's *Bulletin* article [1], which is the first English translation of his talk, is reproduced here.) While Hilbert's talk received a warm reception, it is interesting to learn that at the time not everyone recognized the central role his problems would play within the mathematical community.

> *"The communications made in Sections V and VI, while not necessarily the most valuable mathematically, were yet of the most general interest, and lend themselves best to any general report. The sitting was opened by M. Hilbert's address, in German, on the future problems of mathematics."*

Charlotte Angas Scott (1858-1931), a frequent contributor to the *Bulletin,* is an interesting person in the history of American mathematics and our society. She came from an activist background where her father and grandfather were presidents of colleges. The family, particularly her grandfather, had a history of struggling for social reform, education for the working classes, and fighting against slavery and alcohol consumption. With a similar feisty spirit, in 1885 Scott overcame the gender prejudice of her period to become the first English speaking woman to earn a doctorate in mathematics. Actually, the only woman to precede her with a doctorate in mathematics was Sofia Kovalevskaia. Scott's advisor was Arthur Cayley; although she did her research at the University of Cambridge, the university would not grant her a degree because she was a "she." So, Scott arranged to receive her degree from the University of London. Four years later she accepted a position at Bryn Mawr College. Scott was active in our society; she is one of the founders of the AMS serving on our first council and then as an AMS vice-president in 1905. For more information about this fascinating woman, I recommend Kenschaft's articles [9, 10].

Meet Me in Saint Louis; 1904

As a brief political aside, fearful of Napoleon Bonaparte and the French intentions directed toward the New World, President Thomas Jefferson sent Robert Livingston and then James Monroe to Paris either to purchase land near New Orleans on the lower Mississippi river, or at least obtain a guarantee of free navigation on the river. There was good reason for Jefferson's fear; Napoleon had hoped to use the region around New Orleans as means to supply Haiti, which was to be the center and heart of his empire in the Americas. But faced with a rebellion on Haiti and renewed war with England, Napoleon needed funds and could not spare the troops needed for his American expansion dreams. The result was the Louisiana Purchase of April 30, 1803; a purchase that doubled the size of the new country. The next year, 1804, started the Lewis and Clark expedition from St. Louis.

A century later, the 1903 "Louisiana Purchase Exposition" — described as the St. Louis World's Fair and probably better remembered from the song "Meet me in Saint Louis" — commemorated this important event. The relevance of this fair for mathematics is that in September, 1904, the World's Fair sponsored lectures by mathematicians and others during their International Congress of the Arts and Sciences. Several of the written versions of these talks appeared in the *Bulletin;* three are reproduced in this book. One of the lectures is the Pierpont paper mentioned earlier; it is the first article reproduced in this section. Another is an important address by Poincaré where he describes mathematical notions that are tantalizingly close to special relativity; this is described in a later section. A third delightful paper, which is the fourth and last reproduced in this section, is the written version of Jean Gaston Darboux's [5] address.

Darboux (1842-1917), a trend-setter of the time known through his work on differential geometry, the use of analysis to understand properties of curves and surfaces, and the "Darboux integral" (introduced in an 1870 paper on differential equations), presented his paper "A survey of the development of geometric methods." He knew how to attract attention; after all, how can any mathematician, who appreciates the demands of our field, not be amused by his opening commentary that

"*... in the last period of his life, Lagrange, fatigued by those re-*

*searches in analysis and mechanics which have assured to him
an immortal renown, began to neglect mathematics for chemistry which according to him was becoming as easy as algebra,
..."*

Among other topics, Darboux stressed the consequences of the methods
of Monge[5] and Poncelet as developed by them and their disciples.

*"Modern geometry — this we must claim as its title to distinction
— arrived at the very end of the eighteenth century to contribute
in large measure to the rejuvenation of the whole of mathematical
science by offering a new and productive field for investigation
and especially by proving to us, by means of brilliant successes,
that general methods do not exhaust a science, but that even
in the simplest subject there remains much to be done by an
ingenious and inventive mind."*

More to the point, Darboux includes the quote

*" 'With his geometry,' says Lagrange, speaking of Mange, 'this
devil of a man will make himself immortal.' "*

Beyond describing geometry at this time, Darboux provides some personal aspects of the history. As an example, Poncelet "...gave his efforts
exclusively to the development of the purely geometric germs contained in
the researches of his illustrious predecessor [Monge]." But rather doing what
we do, where we seek funding to support summer research,

*"[m]ade prisoner by the Russians in 1813 at the crossing of the
Dnieper and confined at Saratoff, Poncelet[6] employed the leisure*

---

[5] At the age of 18, in recognition of the city plan he drew for Beaune, Monge (1746-1818)
was appointed to the École Royale du Génie where he encountered Bossut; it was Bossut
who encouraged Monge's mathematical interests. In addition to his partial differential
equations approach, geometry, and other mathematical interests, Monge was interested
in metallurgy, chemistry, physics — and the politics of the day. This included the 1793
Revolution (where he was involved in absolving the royalists of the charge that they mixed
glass with flour intended for the revolutionary armies, and voted for the death of the king)
and his later friendship with, and support of, Napoleon – a fact that eventually caused
both his expulsion from the Institut de France and personal difficulties for the rest of his
life.

[6] Poncelet (1788-1867) was a student of Monge. This incident occurred when, as an
engineer, he took part in Napoleon's 1812 Russian campaign and was left for dead. After
he was released in 1814, Poncelet, one of the founders of modern projective geometry, was
a military engineer and then a professor of *mechanics* at Metz.

of captivity in demonstrating those principles which he developed in the Traité des propriétés projectives des figures, published in 1822, ... And thus Saratoff may be considered the birthplace of modern geometry."

In addition to being an exceptional mathematician, Darboux was known for his teaching, exposition, history of science (e.g., he was a biographer of Poincaré), and even as an administrator. He was honored by being elected to over a hundred scientific societies, and, to see a more concrete form of recognition, when walking around the 18th Arrondissement of Paris look for "Rue Gaston Darboux." Darboux was born in Nimes, and educated at Ecole Normale Supérieure and the Sorbonne. Later, returning to the Sorbonne, he held the chair on higher geometry. For more about him, I suggest Alexander [1] and Eisenhart [6].

Bibliography

[1] Alexander, D. S., Gaston Darboux and the history of complex dynamics, *Historia Mathematica* **22** (1995), 179-185.

[2] Cajori, F., Evolution of criteria of convergence, *Bull of New York Math Soc.* **3** (1893), 1-10.

[3] Chapman, C. H., The theory of transformation groups (Book review), *Bull of New York Math Soc.* **3** (1893), 61-71.

[4] Collidge, J. L., The opportunities for mathematical study in Italy, *BAMS* **11** (1904), 9-17.

[5] Darboux, M. G., A survey of the development of geometric methods, *BAMS* **12**, 517-543

[6] L P Eisenhart, L. P. Darboux's contribution to geometry, *Bull. Amer. Math. Soc.* **24** (1917/18), 227-237.

[7] Hilbert, D., Mathematical Problems, *BAMS* **8**, (1902), 437-479.

[8] Johnson, W., Octonary Numeration, *Bull of New York Math Soc.* **1** (1891), 1-6.

[9] Kenschaft, P., Charlotte Angas Scott, 1858-1931, College Mathematics Journal **18** (1987), 98-110

[10] Kenschaft, P., Charlotte Angas Scott (1858-1931), in *Women of Mathematics; A Biographic Sourcebook.* L. S. Grinstein and P. J. Campbell eds., Greenwood Press, Inc., Westport, CT. 1987

[11] McClintock, E., The past and future of the Society, *BAMS* **2** (1895), 85-94.

[12] Parshall, K. H., and D. E. Rowe, *The emergence of the American Mathematical Research Community 1876 - 1900: J. J. Sylvester, Felix Klein, and E. H. Moore*, American Mathematics Society, Providence, 1994.

[13] Pierpont, J., The History of Mathematics in the Nineteenth Century, *BAMS* **11** (1904), 136-159. (From BAMS article in Jan. 2000.)

[14] Pierpont, J., Mathematical rigor, past and present, *BAMS* **34** (1928), 25-53.

[15] Scott, C. A., Paris Congress of Mathematicians, *BAMS* **7** (1900), 57-79

THE HISTORY OF MATHEMATICS IN THE NINETEENTH CENTURY.

*ADDRESS DELIVERED BEFORE THE DEPARTMENT OF MATHE-
MATICS OF THE INTERNATIONAL CONGRESS OF ARTS
AND SCIENCE, ST. LOUIS, SEPTEMBER 20, 1904.*

BY PROFESSOR JAMES PIERPONT.

The extraordinary development of mathematics in the last century is quite unparalleled in the long history of this most ancient of sciences. Not only have those branches of mathematics which were taken over from the eighteenth century steadily grown but entirely new ones have sprung up in almost bewildering profusion, and many of these have promptly assumed proportions of vast extent.

As it is obviously impossible to trace in the short time allotted to me the history of mathematics in the nineteenth century, even in merest outline, I shall restrict myself to the consideration of some of its leading theories.

Theory of Functions of a Complex Variable.

Without doubt one of the most characteristic features of mathematics in the last century is the systematic and universal use of the complex variable. Most of the great mathematical theories received invaluable aid from it, and many owe to it their very existence. What would the theory of differential equations or elliptic functions be to-day without it, and is it probable that Poncelet, Steiner, Chasles, and von Staudt would have developed synthetic geometry with such elegance and perfection without its powerful stimulus?

The necessities of elementary algebra kept complex numbers persistently before the eyes of every mathematician. In the eighteenth century the more daring, as Euler and Lagrange, used them sparingly; in general one avoided them when possible. Three events, however, early in the nineteenth century changed the attitude of mathematicians toward this mysterious guest. In 1813–14 Argand published his geometric interpretation of complex numbers. In 1824 came the discovery by Abel of the imaginary period of the elliptic function. Finally Gauss in his

Reprinted from Bull. Amer. Math. Soc. 11 (December 1904), 136–159

second memoir on biquadratic residues (1832) proclaims them a legitimate and necessary element of analysis.

The theory of functions of a complex variable may be said to have had its birth when Cauchy discovered his integral theorem

$$\int_c f(x)dx = 0,$$

published in 1825. In a long series of publications beginning with the Cours d'analyse, 1821, Cauchy gradually developed his theory of functions and applied it to problems of the most diverse nature ; e. g., existence theorems for implicit functions and the solutions of certain differential equations, the development of functions in infinite series and products, and the periods of integrals of one-valued and many-valued functions.

Meanwhile Germany is not idle ; Weierstrass and Riemann develop Cauchy's theory along two distinct and original paths. Weierstrass starts with an explicit analytic expression, a power series, and defines his function as the totality of its analytical continuations. No appeal is made to geometric intuition, his entire theory is strictly arithmetical. Riemann, growing up under Gauss and Dirichlet, not only relies largely on geometric intuition, but also does not hesitate to impress mathematical physics into his service. Two noteworthy features of his theory are the many-leaved surfaces named after him, and the extensive use of conformal representation.

The history of functions as first developed is largely a theory of algebraic functions and their integrals. A general theory of functions is only slowly evolved. For a long time the methods of Cauchy, Riemann, and Weierstrass were cultivated along distinct lines by their respective pupils. The schools of Cauchy and Riemann were the first to coalesce. The entire rigor which has recently been imparted to their methods has removed all reason for founding, as Weierstrass and his school have urged, the theory of functions on a single algorithm, viz., the power series. We may therefore say that at the close of the century there is only one theory of functions, in which the ideas of its three great creators are harmoniously united.

Let us note briefly some of its lines of advance. Weierstrass early observed that an analytic expression might represent different analytic functions in different regions. Associated with this is the phenomenon of natural boundaries. The question therefore arose as to what is the most general domain of definition of an analytic function.   Runge has shown that any connected region may serve this purpose.   An important line of investigation relates to the analytic expression of a function by means of infinite series, products, and fractions.   Here may be mentioned Weierstrass's discovery of prime factors; the theorems of Mittag-Leffler and Hilbert; Poincaré's uniformization of algebraic and analytic functions by means of a third variable, and the work of Stieltjes, Pincherle, Padé, and Van Vleck on infinite fractions.

Since an analytic function is determined by a single power series, which in general has a finite circle of convergence, two problems present themselves: to determine 1) the singular points of the analytic function so defined, and 2) an analytic expression valid for its whole domain of definition.   The celebrated memoir of Hadamard inaugurated a long series of investigations on the first problem; while Mittag-Leffler's star theorem is the most important result yet obtained relating to the second.

Another line of investigation relates to the work of Poincaré, Borel, Stieltjes, and others on divergent series.   It is indeed a strange vicissitude of our science that these series, which early in the century were supposed to be banished once and for all from rigorous mathematics, should at its close be knocking at the door for readmission.

Let us finally note an important series of memoirs on integral transcendental functions beginning with Weierstrass, Laguerre, and Poincaré.

Algebraic Functions and Their Integrals.

A branch of the theory of functions has been developed to such an extent that it may be regarded as an independent theory, we mean the theory of algebraic functions and their integrals.   The brilliant discoveries of Abel and Jacobi in the elliptic functions from 1824 to 1829 prepared the way for a similar treatment of the hyperelliptic case.   Here a difficulty of gravest nature was met.   The corresponding integrals have $2p$ linearly independent periods; but, as Jacobi had shown, a one-valued function having more than two periods admits a period as small as we choose.   It therefore looked as if the elliptic functions admitted no further generalization.   Guided by Abel's theorem, Jacobi at last discovered the solution to the difficulty, 1832 ; to get functions analogous to the elliptic functions we must consider functions not of one but of p independent variables, viz., the p independent integrals of the first species. The great problem now before mathematicians, known as Jacobi's problem of inversion, was to extend this aperçu to the case of any algebraic configuration and develop the consequences. The first to take up this immense task were Weierstrass and Riemann, whose results belong to the most brilliant achievements of the century. Among the important notions hereby introduced we note the following : the birational transformation, rank of an algebraic configuration, class invariants, prime functions, the theta and multiply periodic functions in several variables. Of great importance is Riemann's method of proving existence theorems as also his representation of algebraic functions by means of integrals of the second species.

A new direction was given to research in this field by Clebsch, who considered the fundamental algebraic configuration as defining a curve. His aim was to bring about a union of Riemann's ideas and the theory of algebraic curves for their mutual benefit. Clebsch's labors were continued by Brill and Noether ; in their work the transcendental methods of Riemann are placed quite in the background. More recently Klein and his school have sought to unite the transcendental methods of Riemann with the geometric direction begun by Clebsch, making systematic use of homogeneous coördinates and the invariant theory. Noteworthy also is Klein's use of normal curves in $(p - 1)$-way space to represent the given algebraic configuration. Dedekind and Weber, Hensel and Landsberg have made use of the ideal theory with marked success. Many of the difficulties of the older theory, for example the resolution of singularities of the algebraic configuration, are treated with a truly remarkable ease and generality.

In the theory of multiply periodic functions and the general θ functions we mention, besides those of Weierstrass, the researches of Prym, Krazer, Frobenius, Poincaré, and Wirtinger.

Automorphic Functions.

Closely connected with the elliptic functions is a class of functions which has come into great prominence in the last quarter of a century, viz. : the elliptic modular and automorphic functions. Let us consider first the modular functions of which the modulus κ and the absolute invariant J are the simplest types.

The transformation theory of Jacobi gave algebraic relations between such functions in endless number. Hermite, Fuchs, Dedekind and Schwarz are forerunners, but the theory of modular functions as it stands to-day is principally due to Klein and his school. Its goal is briefly stated thus : To determine all subgroups of the linear group

$$x' = \frac{\alpha x + \beta}{\gamma x + \delta},\qquad (1)$$

where $\alpha, \beta, \gamma, \delta$ are integers and $\alpha\delta - \beta\gamma \neq 0$; to determine for each such group associate modular functions and to investigate their relation to one another and especially to J. Important features in this theory are the congruence groups of (1) ; the fundamental polygon belonging to a given subgroup, and its use as substitute for a Riemann surface ; the principle of reflection on a circle ; the modular forms.

The theory of automorphic functions is due to Klein and Poincaré. It is a generalization of the modular functions ; the coefficients in (1) being any real or imaginary numbers, with non-vanishing determinant, such that the group is discontinuous. Both authors have recourse to non-euclidean geometry to interpret the substitutions (1). Their manner of showing the existence of functions belonging to a given group is quite different. Poincaré by a brilliant stroke of genius actually writes down their arithmetic expressions in terms of his celebrated θ series. Klein employs the existence methods of Riemann. The relation of automorphic functions to differential equations is studied by Poincaré in detail. In particular, he shows that both variables of a linear differential equation with algebraic coëfficients can be expressed uniformly by their means.

Differential Equations.

Let us turn now to another great field of mathematical activity, the theory of differential equations. The introduction of the theory of functions has completely revolutionized this subject. At the beginning of the nineteenth century many important results had indeed been established, particularly by Euler and Lagrange ; but the methods employed were artificial, and broad comprehensive principles were lacking. By various devices one tried to express the solution in terms of the elementary functions and quadratures — a vain attempt; for, as we know now, the goal they strove so laboriously to reach was in general unattainable.

A new epoch began with Cauchy, who by means of his new theory of functions first rigorously established the existence of the solution of certain classes of equations in the vicinity of regular points. He also showed that many of the properties of the elliptic functions might be deduced directly from their differential equations. Ere long, the problem of integrating a differential equation changed its base. Instead of seeking to express its solution in terms of the elementary functions and quadratures, one asked what is the nature of the functions defined by a given equation. To answer this question we must first know what are the singular points of the integral function and how it behaves in their vicinity. The number of memoirs on this fundamental and often difficult question is enormous; but this is not strange if we consider the great variety of interesting and important classes of equations which have to be studied.

One of the first to open up this new path was Fuchs, whose classic memoirs (1866–68) gave the theory of linear differential equations its birth. These equations enjoy a property which renders them particularly accessible, viz., the absence of movable singular points. They may, however, possess points of indetermination, to use Fuchs's terminology, and little progress has been made in this case. Noteworthy in this connection is the introduction by von Koch of infinite determinants, whose importance was first shown by our distinguished countryman, Hill; also the use of divergent series — that invention of the devil, as Abel called them — by Poincaré. A particular class of linear differential equations of great importance is the hypergeometric equation; the results obtained by Gauss, Kummer, Riemann, and Schwarz relating to this equation have had the greatest influence on the development of the general theory. The great extent of the theory of linear differential equations may be estimated when we recall that within its borders it embraces not only almost all the elementary functions, but also the modular and automorphic functions.

Too important to pass over in silence is the subject of algebraic differential equations with uniform solutions. The brilliant researches of Painlevé deserve especial mention.

Another field of great importance, especially in mathematical physics, relates to the determination of the solution of differential equations with assigned boundary conditions. The literature of this subject is enormous ; we may therefore be pardoned if mention is made only of the investigations of Green, Sturm, Liouville, Bôcher, Riemann, Schwarz, C. Neumann, Poincaré, and Picard.

Since 1870 the theory of differential equations has been greatly advanced by Lie's theory of groups. Assuming that an equation or a system of equations admits one or more infinitesimal transformations, Lie has shown how these may be employed to simplify the problem of integration. In many cases they give us exact information how to conduct the solution and upon what system of auxiliary equations the solution depends. One of the most striking illustrations of this is the theory of ordinary linear differential equations which Picard and Vessiot have developed, analogous to Galois's theory for algebraic equations. An interesting result of this theory is a criterion for the solution of such equations by quadratures. As an application we find that Riccati's equation cannot be solved by quadratures. The attempts to effect such a solution of this celebrated equation in the preceding century were therefore necessarily in vain.

A characteristic feature of Lie's theories is the prominence given to the geometrical aspects of the questions involved. Lie thinks in geometrical images, the analytical formulation comes afterwards. Already Monge had shown how much might be gained by geometrizing the problem of integration. Lie has gone much farther in this direction. Besides employing all the geometrical notions of his predecessors extended to n-way space, he has introduced a variety of new conceptions, chief of which are his surface element and contact transformations.

He has also used with great effect Plücker's line geometry and his own sphere geometry in the study of certain types of partial differential equations of the first and second orders which are of great geometrical interest, for example equations whose characteristic curves are lines of curvature, geodesics, etc. Let us close by remarking that Lie's theories not only afford new and valuable points of view for attacking old problems but also give rise to a host of new ones of great interest and importance.

Groups.

We turn now to the second dominant idea of the century, the group concept.

Groups first became objects of study in algebra when Lagrange 1770, Ruffini 1799, and Abel 1826 employed substitution groups with great advantage to their work on the quintic. The enormous importance of groups in algebra was, however, first made clear by Galois, whose theory of the solution of algebraic equations is one of the great achievements of the century. Its influence has stretched far beyond the narrow bounds of algebra.

With an arbitrary but fixed domain of rationality, Galois observed that every algebraic equation has attached to it a certain group of substitutions. The nature of the auxiliary equations required to solve the given equation is completely revealed by an inspection of this group.

Galois's theory showed the importance of determining the sub-groups of a given substitution group, and this problem was studied by Cauchy, Serret, Mathieu, Kirkman and others. The publication of Jordan's great treatise in 1870 is a noteworthy event. It collects and unifies the results of his predecessors and contains an immense amount of new matter.

A new direction was given to the theory of groups by the introduction by Cayley of abstract groups (1854, 1878). The work of Sylow, Hölder, Frobenius, Burnside, Cole, and Miller deserves especial notice.

Another line of researches relates to the determination of the finite groups in the linear group of any number of variables. These groups are important in the theory of linear differential equations with algebraic solutions; in the study of certain geometrical problems, as the points of inflection of a cubic, the 27 lines on a surface of the third order; in crystallography, etc. They also enter prominently in Klein's Formenproblem. An especially important class of finite linear groups are the congruence groups first considered by Galois. Among the laborers in the field of linear groups we note Jordan, Klein, Moore, Maschke, Dickson, Frobenius, and Wiman.

Up to the present we have considered only groups of finite order. About 1870 entirely new ideas, coming from geometry and differential equations, give the theory of groups an unexpected development. Foremost in this field are Lie and Klein.

Lie discovers and gradually perfects his theory of continuous transformation groups and shows their relations to many different branches of mathematics. In 1872 Klein publishes his Erlanger Programm and in 1877 begins his investigations on elliptic modular functions, in which infinite *discontinuous* groups are of primary importance, as we have already seen.   In the now famous Erlanger Programm, Klein asks what is the principle which underlies and unifies the heterogeneous geometric methods then in vogue, as for example, the geometry of the ancients whose figures are rigid and invariable, the modern projective geometry, whose figures are in ceaseless flux, passing from one form to another, the geometries of Plücker and Lie in which the elements of space are no longer points but lines, spheres, or other configurations at pleasure, the geometry of birational transformations, the analysis situs, etc., etc.   Klein finds this answer : In each geometry we have a system of objects and a group which transforms these objects one into another.   We seek the invariants of this group.   In each case it is the abstract group which is essential, and not the concrete objects.   The fundamental role of a group in geometrical research is thus made obvious.   Its importance in the solution of algebraic equation, in the theory of differential equations, in the automorphic functions we have already seen.   The immense theory of algebraic invariants developed by Cayley and Sylvester, Aronhold, Clebsch, Gordan, Hermite, Brioschi, and a host of zealous workers in the middle of the century, also finds its place in the far more general invariant theory of Lie's theory of groups.   The same is true of the theory of surfaces as far as it rests on the theory of differential forms.   In the theory of numbers, groups have many important applications, for example, in the composition of quadratic forms and the cyclotomic bodies.   Finally let us note the relation between hypercomplex numbers and continuous groups discovered by Poincaré.

In resumé, we may thus say that the group concept, hardly noticeable at the beginning of the century, has at its close become one of the fundamental and most fruitful notions in the whole range of our science.

Infinite Aggregates.

Leaving the subject of groups, we consider now briefly another fundamental concept, viz., infinite aggregates.   In the most diverse mathematical investigations we are confronted with such aggregates.   In geometry the conceptions of a curve, surface, regon, frontier, etc., when examined carefully, lead us to a rich variety of aggregates. In analysis they also appear, for example the domain of definition of an analytical function, the points where a function of a real variable ceases to be continuous or to have a differential coefficient, the points where a series of functions ceases to be uniformly convergent, etc.

To say that an aggregate (not necessarily a point aggregate) is infinite is often an important step; but often again only the first step. To penetrate farther into the problem may require us to state *how* infinite. This requires us to make distinctions in infinite aggregates, to discover fruitful principles of classification, and to investigate the properties of such classes.

The honor of having done this belongs to Georg Cantor. The theory of aggregates is for the most part his creation; it has enriched mathematical science with fundamental and far reaching notions and results.

The theory falls into two parts; a theory of aggregates in general, and a theory of point aggregates. In the theory of point aggregates the notion of limiting points gives rise to important classes of aggregates, as discrete, dense, complete, perfect, connected, etc., which are so important in the function theory.

In the general theory two notions are especially important, viz. : the one-to-one correspondence of the elements of two aggregates, and well ordered aggregates. The first leads to cardinal numbers and the idea of enumerable aggregates, the second to transfinite or ordinal numbers.

Two striking results of Cantor's theory may be mentioned : the algebraic and therefore the rational numbers, although everywhere dense, are enumerable ; and secondly, one-way and n-way space have the same cardinal number.

Cantor's theory has already found many applications, especially in the function theory, where it is today an indispensable instrument of research.

Functions of Real Variables. The Critical Movement.

One of the most conspicuous and distinctive features of mathematical thought in the nineteenth century is its critical spirit. Beginning with the calculus, it soon permeates all analysis, and toward the close of the century it overhauls and recasts the foundations of geometry and aspires to further conquests in mechanics and in the immense domains of mathematical physics.

Ushered in with Lagrange and Gauss just at the close of the eighteenth century, the critical movement receives its first decisive impulse from the teachings of Cauchy, who in particular introduces our modern definition of limit and makes it the foundation of the calculus. We must also mention in this connection Abel, Bolzano, and Dirichlet. Especially Abel adopted the reform ideas of Cauchy with enthusiasm and made important contributions in infinite series.

The figure, however, which towers above all others in this movement, whose name has become a synonym of rigor, is Weierstrass. Beginning at the very foundations, he creates an arithmetic of real and complex numbers, assuming the theory of positive integers to be given. The necessity of this is manifest when we recall that until then the simplest properties of radicals and logarithms were utterly devoid of a rigorous foundation; so for example $\sqrt{2}\ \sqrt{5} = \sqrt{10}$, $\log 2 + \log 5 = \log 10$.

Characteristic of the pre-Weierstrassian era is the loose way in which geometrical and other intuitional ideas were employed in the demonstration of analytic theorems. Even Gauss is open to this criticism. The mathematical world received a great shock when Weierstrass showed them an example of a continuous function without a derivative, and Hankel and Cantor by means of their principle of condensation of singularities could construct analytical expressions for functions having in any interval, however small, an infinity of points of oscillation, an infinity of points in which the differential coefficient is altogether indeterminate, or an infinity of points of discontinuity. Another rude surprise was Cantor's discovery of the one-to-one correspondence between the points of a unit segment and a unit square, followed up by Peano's example of a space filling curve.

These examples and many others made it very clear that the idea of a curve, a surface region, motion, etc., instead of being clear and simple, were extremely vague and complex. Until these notions had been cleared up, their admission in the demonstration of an analytical theorem was therefore not to be tolerated. On a purely arithmetical basis, with no appeal to our intuition, Weierstrass develops his stately theory of functions, which culminates in the theory of abelian and multiply periodic functions.

But the notion of rigor is relative and depends on what we are willing to assume either tacitly or explicitly. As we observed, Gauss, whose rigor was the admiration of his contemporaries, freely admitted geometrical notions. This Weierstrass would criticise. On the other hand Weierstrass has committed a grave oversight : he nowhere shows that his definitions relative to the numbers he introduces do not involve mutual contradictions. If he replied that such contradictions would involve contradictions in the theory of positive integers, one might ask what assurance have we that such contradictions may not actually exist? A flourishing young school of mathematical logic has recently grown up in Italy under the influence of Peano. They have investigated with marked success the foundations of analysis and geometry, and have in particular endeavored to show the non-contradictoriness of the axioms of our number system by making them depend on the axioms of logic, which axioms we must admit in order to reason at all.

The critical spirit which in the first half of the century was to be found in the writings of only a few of the foremost mathematicians has in the last quarter of the century become almost universal, at least in analysis. A searching examination of the foundations of arithmetic and the calculus has brought to light the insufficiency of much of the reasoning formerly considered as conclusive. It became necessary to build up these subjects anew. The theory of irrational numbers invented by Weierstrass has been supplanted by the more flexible theories of Dedekind and Cantor. Stolz has given us a systematic and rigorous treatment of arithmetic. The calculus has been completely overhauled and arithmetized by Thomae, Harnack, Peano, Stolz, Jordan, and Vallée-Poussin.

Leaving the calculus, let us notice briefly the theory of functions of real variables. The line of demarcation between these two subjects is extremely arbitrary. We might properly place in the latter all those finer and deeper questions relating to the number system, the study of the curve, surface, and other geometrical notions, the peculiarities that functions present with reference to discontinuity, oscillation, differentiation, and integration, as well as a very extensive class of investigations whose object is the greatest possible extension of the processes, concepts, and results of the calculus. Among the many not yet mentioned who have made important contributions to this subject we note: Fourier, Riemann, Stokes, Dini, Tannery, Pringsheim, Arzelà, Osgood, Brodén, Ascoli, Borel, Baire, Köpcke, Hölder, Volterra, and Lebesgue.

Closely related with the differential calculus is the calculus of variations; in the former the variables are given infinitesimal variations, in the latter the functions. Developed in a purely formal manner by Jacobi, Hamilton, Clebsch and others in the first part of the century, a new epoch began with Weierstrass who, having subjected the labors of his predecessors to an annihilating criticism, placed the theory on a new and secure foundation, and so opened the path for further research by Schwarz, A. Mayer, Scheeffer, von Escherich, Kneser, Osgood, Bolza, Kobb, Zermelo and others. At the very close of the century Hilbert has given the theory a fresh impulse by the introduction of new and powerful methods which enable us in certain cases to neglect the second variation and to simplify the consideration of the first.

Theory of Numbers. Algebraic Bodies.

The theory of numbers as left by Fermat, Euler and Legendre was for the most part concerned with the solution of diophantine equations, i. e., given an equation $f(x, y, z, \cdots) = 0$, whose coefficients are integers, to find all rational, and especially all integral solutions. In this problem Lagrange had shown the importance of considering the theory of forms. A new era begins with the appearance of Gauss's Disquisitiones Arithmeticæ in 1801. This great work is remarkable for three things; 1) The notion of divisibility in the form of congruences is shown to be an instrument of wonderful power; 2) the diophantine problem is thrown in the background and the theory of forms is given a dominant role; 3) the introduction of algebraic numbers, viz., the roots of unity.

The theory of forms has been further developed along the lines of the Disquisitiones by Dirichlet, Eisenstein, Hermite, H. J. S. Smith, and Minkowski.

Another part of the theory of numbers also goes back to Gauss, viz., algebraic numerical bodies. The law of reciprocity of quadratic residues, one of the gems of the higher arithmetic, was first rigorously proved by Gauss. His attempts to extend this theorem to cubic and biquadratic residues showed that the elegant simplicity which prevailed in quadratic residues was altogether missing in these higher residues until one passed from the domain of real integers to the domain formed of the third and fourth roots of unity. In these domains, as

Gauss remarked, algebraic integers have essentially the same properties as ordinary integers. Further exploration in this new and promising field by Jacobi, Eisenstein, and others soon brought to light the fact that already in the domain formed of the 23d roots of unity the laws of divisibility were altogether different from those of ordinary integers ; in particular a number could be expressed as the product of prime factors in more than one way. Further progress in this direction was therefore apparently impossible.

It is Kummer's immortal achievement to have made further progress possible by the invention of his ideals. These he applied to Fermat's celebrated last theorem and the law of reciprocity of higher residues.

The next step in this direction was taken by Dedekind and Kronecker, who developed the ideal theory for any algebraic domain. So arose the theory of algebraic numerical bodies which has come into such prominence in the last decades of the century through the researches of Hensel, Hurwitz, Minkowski, Weber, and above all Hilbert.

Kronecker has gone farther ; in his classic Grundzüge he has shown that similar ideas and methods enable us to develop a theory of algebraic bodies in any number of variables. The notion of divisibility, so important in the preceding theories, is generalized by Kronecker still farther in the shape of his system of moduli.

Another noteworthy field of research opened up by Kronecker is the relation between binary quadratic forms with negative determinant and complex multiplication of elliptic functions. H. J. S. Smith, Gierster, Hurwitz, and especially Weber have made important contributions.

A method of great power in certain investigations has been created by Minkowski which he calls the Geometrie der Zahlen. Introducing a generalization of the distance function, he is led to the conception of a fundamental body (Aichkörper). Minkowski shows that every fundamental body is nowhere concave and conversely to each such body belongs a distance function. A theorem of great importance is now the following : The minimum value which each distance function has at the lattice points is not greater than a certain number depending on the function chosen.

We wish finally to mention a line of investigation which makes use of the infinitesimal calculus and even the theory of

functions.   Here belong the brilliant researches of Dirichlet relating to the number of classes of binary forms for a given determinant, the number of primes in a given arithmetic progression ; and Riemann's remarkable memoir on the number of
primes in a given interval.

On this analytical side of the theory of numbers we notice
also the researches of Mertens, Weber, von Mangoldt, and
Hadamard.

Projective Geometry.

The tendencies of the eighteenth century were predominantly
analytic.   Mathematicians were absorbed for the most part
in developing the wonderful instrument of the calculus with
its countless applications.   Geometry made relatively little
progress.   A new era begins with Monge.   His numerous and
valuable contributions to analytic, descriptive, and differential
geometry and especially his brilliant and inspiring lectures at
the Ecole polytechnique (1795–1809) put fresh life in geometry and prepared it for a new and glorious development in the
nineteenth century.

When one passes in review the great achievements which
have made the nineteenth century memorable in the annals of
our science, certainly projective geometry will occupy a foremost place.   Pascal, De la Hire, Monge, and Carnot are forerunners, but Poncelet, a pupil of Monge, is its real creator.
The appearance of his Traité des propriétés projectives des
figures in 1822 gives modern geometry its birth.   In it we
find the line at infinity, the introduction of imaginaries, the
circular points at infinity, polar reciprocation, a discussion of
homology, the systematic use of projection, section, and anharmonic ratio.

While the countrymen of Poncelet, especially Chasles, do
not fail to make numerous and valuable contributions to the
new geometry, the next great steps in advance are made on
German soil.   In 1827 Möbius publishes the Barycentrische
Calcul ; Plücker's Analytisch-geometrische Entwickelungen appears in 1828–31 ; and Steiner's Systematische Entwickelung
der Abhängigkeit geometrischer Gestalten von einander in 1832.
In the ten years which embrace the publication of these immortal works of Poncelet, Plücker, and Steiner, geometry has
made more real progress than in the 2,000 years which had
elapsed since the time of Apollonius.   The ideas which had

been slowly taking shape since the time of Descartes suddenly crystallized and almost overwhelmed geometry with an abundance of new ideas and principles.

To Möbius we owe the introduction of homogeneous coördinates, and the far reaching conception of geometric transformation including collineation and duality as special cases. To Plücker we owe the use of the abbreviated notation which permits us to study the properties of geometric figures without intervention of the coördinates, the introduction of line and plane coördinates and the notion of generalized space elements. Steiner, who has been called the greatest geometer since Apollonius, besides enriching geometry in countless ways, was the first to employ systematically the method of generating geometric figures by means of projective pencils.

Other noteworthy works belonging to this period are Plücker's System der analytischen Geometrie (1835) and Chasles's classic Aperçu (1837).

Already at this stage we notice a bifurcation in geometric methods. Steiner and Chasles become eloquent champions of the synthetic school of geometry, while Plücker and later Hesse and Cayley are leaders in the analytic movement. The astonishing fruitfulness and beauty of synthetic methods threatened for a short time to drive the analytic school out of existence. The tendency of the synthetic school was to banish more and more metrical methods. In effecting this the anharmonic ratio became constantly more prominent. To define this fundamental ratio without reference to measurement and so to free projective geometry from the galling bondage of metric relations was thus a problem of fundamental importance. The glory of this achievement, which has, as we shall see, a far wider significance, belongs to von Staudt. Another equally important contribution of von Staudt to synthetic geometry is his theory of imaginaries. Poncelet, Steiner, Chasles operate with imaginary elements as if they were real. Their only justification is recourse to the so-called principle of continuity or to some other equally vague principle. Von Staudt gives this theory a rigorous foundation, defining the imaginary points, lines and planes by means of involutions without ordinal elements.

The next great advance made is the advent of the theory of algebraic invariants. Since projective geometry is the study of those properties of geometric figures which remain unaltered by projective transformations, and since the theory of invariants is the study of those forms which remain unaltered (except possibly for a numerical factor) by the group of linear substitutions, these two subjects are inseparably related and in many respects only different aspects of the same thing. It is no wonder then that geometers speedily applied the new theory of invariants to geometrical problems. Among the pioneers in this direction were Cayley, Salmon, Aronhold, Hesse, and especially Clebsch.

Finally we must mention the introduction of the line as a space element. Forerunners are Grassmann, 1844, and Cayley, 1859, but Plücker in his memoirs of 1865 and his work Neue Geometrie des Raumes (1868–69), was the first to show its great value by studying complexes of the first and second order and calling attention to their application to mechanics and optics.

The most important advance over Plücker has been made by Klein who takes as coördinates six line complexes in involution. Klein also observed that line geometry may be regarded as a point geometry on a quadric in five-way space. Other laborers in this field are Clebsch, Reye, Study, Segre, Sturm, and Koenigs.

Differential Geometry.

During the first quarter of the century this important branch of geometry was cultivated chiefly by the French. Monge and his school study with great success the generation of surfaces in various ways, the properties of envelopes, evolutes, lines of curvature, asymptotic lines, skew curves, orthogonal systems, and especially the relation between the surface theory and partial differential equations.

The appearance of Gauss's Disquisitiones generales circa superficies curvas in 1828 marks a new epoch. Its wealth of new ideas has furnished material for countless memoirs and given geometry a new direction. We find here the parametric representation of a surface, the introduction of curvilinear coördinates, the notion of spherical image, the gaussian measure of curvature, and a study of geodesics. But by far the most important contributions that Gauss makes in this work are the consideration of a surface as a flexible inextensible film or membrane and the importance given quadratic differential forms.

We consider now some of the lines along which differential geometry has advanced. The most important is perhaps the theory of differential quadratic forms with their associate invariants and parameters. We mention here Lamé, Beltrami, Mainardi, Codazzi, Christoffel, Weingarten, and Maschke.

An especially beautiful application of this theory is the immense subject of applicability and deformation of surfaces in which Minding, Bauer, Beltrami, Weingarten, and Voss have made important contributions.

Intimately related with the theory of applicability of two surfaces is the theory of surfaces of constant curvature which play so important a part in non-euclidean geometry. We mention here the work of Minding, Bonnet, Beltrami, Dini, Bäcklund, and Lie.

The theory of rectilinear congruences has also been the subject of important researches from the standpoint of differential geometry. First studied by Monge as a system of normals to a surface and then in connection with optics by Malus, Dupin, and Hamilton, the general theory has since been developed by Kummer, Ribaucour, Guichard, Darboux, Voss, and Weingarten. An important application of this theory is the infinitesimal deformation of a surface.

Minimum surfaces have been studied by Monge, Bonnet and Enneper. The subject owes its present extensive development principally to Weierstrass, Riemann, Schwarz, and Lie. In it we find harmoniously united the theory of surfaces, the theory of functions, the calculus of variations, the theory of groups, and mathematical physics.

Another extensive division of differential geometry is the theory of orthogonal systems, of such importance in physics. We note especially the investigations of Dupin, Jacobi, Lamé, Darboux, Combescure, and Bianchi.

We have already mentioned the intimate relation between differential geometry and differential equations developed by Monge and Lie. Among the workers in this fruitful field Darboux deserves especial mention.

One of the most original and interesting contributions to geometry in the last decades of the century is Lie's sphere geometry. As a brilliant application of it to differential geometry we may mention the relation discovered by Lie between asymptotic lines and lines of curvature of a surface. The subject of sphere geometry has been developed also by Darboux, Reye, Laguerre, Loria, P. F. Smith, and E. Müller.

Other Branches of Geometry.

Under this head we group a number of subjects too important to pass over in silence, yet which cannot be considered at length for lack of time.

In the first place is the immense subject of algebraic curves and surfaces. Adequately to develop all the important and elegant properties of curves and surfaces of the second order alone, would require a bulky volume. In this line of ideas would follow curves and surfaces of higher order and class. Their theory is far less complete, but this lack it amply makes good by offering an almost bewildering variety of configurations to classify and explore. No single geometer has contributed more to this subject than Cayley.

A theory of great importance is the geometry on a curve or surface inaugurated by Clebsch in 1863. Expressing the coördinates of a plane cubic by means of elliptic functions and employing their addition theorems, he deduced with hardly any calculation Steiner's theorem relating to the inscribed polygons and various theorems concerning conics touching the curve. Encouraged by such successes Clebsch proposed to make use of Riemann's theory of abelian functions in the study of algebraic curves of any order. The most important result was a new classification of such curves. Instead of the linear transformation Clebsch, in harmony with Riemann's ideas, employs the birational transformation as a principle of classification. From this standpoint we ask what are the properties of algebraic curves which remain invariant for such transformation.

Brill and Noether follow Clebsch. Their method is, however, algebraical and rests on their celebrated residual theorem, which in their hands takes the place of Abel's theorem. We mention further the investigations of Castelnuovo, Weber, Krauss and Segre. An important division of this subject is the theory of correspondences. First studied by Chasles for curves of deficiency 0 in 1864, Cayley and immediately after Brill extended the theory to the case of any p. The most important advance made in later years has been made by Hurwitz, who considers the totality of possible correspondences on an algebraic curve, making use of the abelian integrals of the first species.

Alongside the geometry on a curve is the vastly more difficult and complicated geometry on a surface, or more generally, on any algebraic spread in n-way space. Starting from a remark of Clebsch 1868, Noether made the first great step in his famous memoirs of 1868–74.   Further progress has been due to the French and Italian mathematicians.   Picard, Poincaré, and Humbert make use of transcendental methods in which figure prominently double integrals which remain finite on the surface and single integrals of total differentials.   On the other hand Enriques and Castelnuovo have attacked the subject from a more algebraic-geometric standpoint by means of linear systems of algebraic curves on the surface.

The first invariants of a surface were discovered by Clebsch and Noether; still others have been found by Castelnuovo and Enriques in connection with irregular surfaces.

Leaving this subject let us consider briefly the geometry of n dimensions.   A characteristic of nineteenth century mathematics is the generality of its methods and results.   When such has been impossible with the elements in hand, fresh ones have been invented; witness the introduction of imaginary numbers in algebra and the function theory, the ideals of Kummer in the theory of numbers, the line and plane at infinity in projective geometry.   The benefit that analysis derived from geometry was too great not to tempt mathematicians to free the latter from the narrow limits of three dimensions and so give it the generality that the former has long enjoyed.   The first pioneer in this abstract field was Grassmann (1844); we must, however, consider Cayley as the real founder of n-dimensional geometry (1869).   Notable contributions have been made by the Italian school, Veronese, Segre, and others.

Non-Euclidean Geometry.

Each century takes over as a heritage from its predecessors a number of problems whose solution previous generations of mathematicians have arduously but vainly sought.   It is a signal achievement of the nineteenth century to have triumphed over some of the most celebrated of these problems.

The most ancient of them is the quadrature of the circle, which already appears in our oldest mathematical document, the Papyrus Rhind, B. C. 2000.   Its impossibility was finally shown by Lindemann, 1882.

Another famous problem relates to the solution of the quintic, which had engaged the attention of mathematicians since the middle of the sixteenth century.   The impossibility of expressing its roots by radicals was finally shown by the youthful Abel (1824), while Hermite and Kronecker (1858) showed how they might be expressed by the elliptic modular functions, and Klein (1875) by means of the icosahedral irrationality.

But of all problems which have come down from the past by far the most celebrated and important relates to Euclid's parallel axiom. Its solution has profoundly affected our views of space and given rise to questions even deeper and more far reaching, which embrace the entire foundation of geometry and our space conception. Let us pass in rapid review the principal events of this great movement. Wallis in the seventeenth, Saccheri, Lambert, and Legendre in the eighteenth are the first to make any noteworthy progress before the nineteenth century. The really profound investigations of Saccheri and Lambert strangely enough were entirely overlooked by later writers and have only recently come to light.

In the nineteenth century non-euclidean geometry develops along four directions which roughly follow each other chronologically. Let us consider them in order.

The naive-synthetic direction. — The methods employed are similar to those of Euclid. His axioms are assumed with the exception of the parallel axiom ; the resulting geometry is what is now called hyperbolic or Lobachevsky's geometry. Its principal properties are deduced ; in particular its trigonometry, which is shown to be that of a sphere with imaginary radius, as Lambert had divined. As a specific result of these investigations the long-debated question relating to the independence of the parallel axiom was finally settled. The great names in this group are Lobachevsky, Bolyai, and Gauss. The first publications of Lobachevsky are his Exposition succincte des principes de la géométrie (1829) and the Geometrische Untersuchungen (1840). Bolyai's Appendix was published in 1832. As to the extent of Gauss's investigations we can only judge from scattered remarks in private letters and his reviews of books relating to the parallel axiom. His dread of the Geschrei der Böotier, i. e., the followers of Kant, prevented him from publishing his extensive speculations.

The metric-differential direction. — This is inaugurated by three great memoirs by Riemann, Helmholtz, and Beltrami, all published in the same year, 1868.

Beltrami, making use of results of Gauss and Minding, relating to the applicability of two surfaces, shows that the hyperbolic geometry of a plane may be interpreted on a surface of constant negative curvature, the pseudosphere. By means of this discovery the purely logical and hypothetical system of Lobachevsky and Bolyai takes on a form as concrete and tangible as the geometry of a plane.

The work of Riemann is as original as profound. He considers space as an n-dimensional continuous numerical multiplicity which is distinguished from the infinity of other such multiplicities by certain well-defined characters. Chief of them are, 1) the quadratic differential expression which defines the length of an element of arc, and 2) a property relative to the displacements of this multiplicity about a point. There are an infinity of space multiplicities which satisfy Riemann's axioms. By extending Gauss's definition of the curvature k of a surface at a point to curvature of space at a point by considering the geodesic surfaces passing through that point, Riemann finds that all these spaces fall into three classes according as k is equal to, greater, or less than 0. For $n = 3$ and $k = 0$ we have euclidean space; when $k < 0$ we have the space found by Gauss, Lobachevsky and Bolyai; when $k > 0$ we have the space first considered in the long forgotten writings of Saccheri and Lambert, in which the right line is finite.

Helmholtz like Riemann considers space as a numerical multiplicity. To further characterize it Helmholtz makes use of the notions of rigid bodies and free mobility. His work has been revised and materially extended by Lie from the standpoint of the theory of groups.

In the present category, as also in the following one, belong important papers by Killing.

The projective direction. — We have already noticed the efforts of the synthetic school to express metric properties by means of projective relations. In this the circular points at infinity were especially serviceable. An immense step in this direction was taken by Laguerre who showed, 1853, that all angles might be expressed as anharmonic ratios with reference to these points, i. e., with reference to a certain fixed conic. The next advance is made by Cayley in his famous sixth memoir on quantics, 1859. Taking any fixed conic (or quadric, for space) which he calls the absolute, Cayley introduces two expressions depending on the anharmonic ratio with reference to the absolute. When this degenerates into the circular points at infinity, these expressions go over into the ordinary expressions for the distance between two points and the angle between two lines.   Thus all metric relations may be considered as projective relations with respect to the absolute.   Cayley does not seem to be aware of the relation of his work to non-euclidean geometry.   This was discovered by Klein (1871).   In fact, according to the nature of the absolute various geometries are possible ; among these are precisely the three already mentioned. Klein has made many important contributions to non-euclidean geometry.   We mention his modification of von Staudt's definition of anharmonic ratio so as to be independent of the parallel axiom, his discovery of the two forms of Riemann's space, and finally his contributions to a class of geometries first noticed by Clifford, which are characterized by the fact that only certain of its possible displacements can affect space as a whole.

As a result of all these investigations both in the projective, as also in the metric differential direction we are led irresistibly to the same conclusion, viz.: The facts of experience can be explained by all three geometries when the constant k is taken small enough.   It is therefore merely a question of convenience whether we adopt the parabolic, hyperbolic, or elliptic geometry.

The critical synthetic direction represents a return to the old synthetic methods of Euclid, Lobachevsky and Bolyai with the added feature of a refined and exacting logic.   Its principal study is not of non-euclidean but of euclidean geometry.   Its aim is to establish a system of axioms for our ordinary space which are complete, compatible, and irreducible.   The fundamental terms, point, line, plane, between, congruent, etc., are introduced as abstract marks whose properties are determined by inter-relations in the form of axioms.   Geometric intuition has no place in this order of ideas which regards geometry as a mere division of pure logic.   The efforts of this school have already been crowned with eminent success, and much may be expected from it in the future.   Its leaders are Peano, Veronese, Pieri, Padoa, Burali-Forti, and Levi-Civita in Italy, Hilbert in Germany, Moore in America, and Russell in England.

Closing at this point our hasty and imperfect survey of mathematics in the last century let us endeavor to sum up its main characteristics.   What strikes us at once is its colossal proportions and rapid growth in nearly all directions, the great variety of its branches, the generality and complexity of its methods, an inexhaustible creative imagination, the fearless introduction and employment of ideal elements, and an appreciation for a refined and logical development of all its parts.

We who stand on the threshold of a new century can look back on an era of unparalleled progress. Looking into the future an equally bright prospect greets our eyes; on all sides fruitful fields of research invite our labor and promise easy and rich returns. Surely this is the golden age of mathematics!

OUTER ISLAND,
September, 1904.

BULLETIN OF THE

NEW YORK MATHEMATICAL SOCIETY

EVOLUTION OF CRITERIA OF CONVERGENCE.

BY PROF. FLORIAN CAJORI.

THE expressions convergent and divergent series were used for the first time in 1668 by James Gregory. Newton and Leibniz felt the necessity of inquiring into the convergence of infinite series, but they had no proper criteria, excepting a test advanced by Leibniz for alternating series. By Euler and his contemporaries the *formal* treatment of series was greatly extended, while the necessity for determining the convergence was generally lost sight of. To be sure, it was Euler who first observed the semi-convergence of a series. He, moreover, remarked that great care should be exercised in the summation of divergent series. But this warning was not taken so seriously by him as it would be by a modern writer, for in the very same article * in which it occurs Euler did not hesitate to write

$$\ldots + \frac{1}{n^2} + \frac{1}{n} + 1 + n + n^2 + \ldots = 0,$$

simply because

$$n + n^2 + \ldots = \frac{n}{1-n} \; ; \; 1 + \frac{1}{n} + \frac{1}{n^2} + \ldots = \frac{n}{n-1}.$$

The facts are that Euler reached some very pretty results in infinite series, now well known, and also some very absurd results, now quite forgotten. Protests were made by Nicolaus Bernoulli and Varignon against the prevailing reckless use of series : isolated attempts at establishing criteria of convergence are on record : but the dominating sentiment of the age frowned down any proposition which would put limitations upon operations with series. The faults of this period found their culmination in the Combinatorial School in Germany, which has now passed into deserved oblivion.

I. *Special Criteria.* In the progress of mathematics, truth

---

* *Comm. Petrop.*, vol. 11. p. 116.

Reprinted from Bull. New York Math. Soc. **2** (1892–03), 1–10

and error cannot go together long. Doubtful or plainly absurd results obtained from infinite series stimulated profounder inquiries into the validity of operations with them. Their *actual contents* came to be the primary, *form* a secondary consideration. The first important and strictly rigorous investigation of series was made by Gauss,[*] when he discussed the series designated by $F(\alpha, \beta, \gamma, x)$. He investigates the ratio of two successive terms, compares series with geometric progressions, and deduces a general criterion for series of positive terms, the ratio of successive terms of which can be expressed in a certain form. The deduction of this criterion is laborious, but it settles the question of convergence in every case which it is intended to cover, and thus bears the stamp of generality, so characteristic of Gauss's writings. Owing to the strangeness of treatment and unusual rigor, Gauss's paper excited little interest among the mathematicians of that time.

More fortunate in reaching the public was Baron Cauchy whose *Analyse Algébrique* of 1821 contains a rigorous treatment of series. All series whose sum does not approach a fixed limit as the number of terms increases indefinitely, are called divergent. Like Gauss, he institutes comparisons with geometric series and finds that series with positive terms are convergent or not, according as $(u_n)^{\frac{1}{n}}$ or u_{n+1}/u_n is ultimately less or greater than unity. To reach cases where these expressions become unity and fail, Cauchy proved (1) that if in $u_0 + u_1 + \ldots$ each term is smaller than its preceding, it is always convergent when $u_0 + 2u_1 + 4u_3 + 8u_7 + \ldots$ is convergent, and only then; (2) that if $Lu_n / L\frac{1}{n}$[†] converges toward a finite limit greater than one, the series is convergent, but divergent when that limit is less than one. He shows that series with partly negative terms are convergent when the absolute values of the terms converge, and then deduces Leibniz's test for alternating series.

The most outspoken critic of the old methods in series was Abel. His letter to his friend Holmboe (June, 1826) contains severe criticisms. It is very interesting reading, even to modern students. Abel also pointed out in *Crelle's Journal*, vol. 3, the error in an article by Olivier, who pretended to have found the following extremely simple, general criteria for a series with positive terms : Σu_n diverges if the limit of nu_n is not zero and converges if it equals zero. Abel showed that the second part of this is incorrect. Among the posthumous papers of Abel [‡] it is demonstrated that a series having

---

[*] " Disquisitiones generales circa seriem infinitam ".. , *Werke*, vol. 3.
[†] L denotes characteristic of logarithm in any system.
[‡] *Oeuvres*, vol. 2, p. 197.

the general term $\dfrac{1}{n \lg n \lg^2 n \ldots \lg^{m-1} n}$ is divergent,[*] but one

having the general term $\dfrac{1}{n \lg n \lg^2 n \ldots (\lg^m n)^{1+a}}$ converges

if $\alpha > 0$. In the same paper he deduces logarithmic tests in anticipation of De Morgan and Bertrand.

The researches of Cauchy and Abel caused a considerable stir. We are told that after a scientific meeting in which Cauchy had presented his first researches on series, Laplace hastened home and remained there in seclusion until he had examined the series in his *Mécanique céleste*. Luckily, every one was found to be convergent! We must not conclude, however, that the new ideas at once displaced the old. On the contrary, the new views were generally accepted only after a severe and long struggle. As late as 1844 De Morgan began a paper on "Divergent Series" in this style: "I believe it will be generally admitted that the heading of this paper describes the only subject yet remaining, of an elementary character, on which a serious schism exists among mathematicians as to absolute correctness or incorrectness of results."[†] Some mathematicians, for instance Poisson, promptly rejected infinitely diverging series, but seemed to employ with confidence finitely diverging series; they appeared content to equate $\frac{1}{2}$ to $1 - 1 + 1 - \ldots$, if this series be regarded as the limit of a convergent series $1 - g + g^2 - \ldots$ Difficult questions of this nature arose in the study of certain trigonometric series, particularly that of Fourier, upon which much light was thrown by the researches of Dirichlet.

First in time, in the evolution of more delicate criteria of convergence and divergence come the researches of Raabe[‡] who starts out with the following theorem previously given by Cauchy in 1827, and subsequently used by other writers: "If $\varphi(x)$ be a function which becomes zero when x increases indefinitely and which has for all values of x beween a and ∞, always finite values of the same sign, then $\Sigma^\infty \varphi(a + n)$ is convergent or divergent, according as $\displaystyle\int_a^\infty \varphi(x)\,dx$ is finite or infinite." Raabe then shows that the series with the general term $e^{-x(1 + \frac{1}{2} + \cdots + \frac{1}{n})}$ is convergent when $x > 1$ and divergent when $x \gtrless 1$; and thence deduces the theorem that a series Σu_n is convergent when, in the expression $\lim \left(\dfrac{u_{n+1}}{u_n}\right)^n$ $= e^{-x}$ or in the equivalent expression (a test bearing the

---

[*] The index is here used as in the calculus of operations, *e.g.*, $\lg^2 n \equiv \lg \lg n$.

[†] *Trans. Cambr. Philos. Soc.*, vcl. 8, pt. ii.

[‡] *Zeitschrift für Physik und Math.*, vol. 10 (1832).

name of the author) $\lim n \left(\dfrac{u_n}{u_{n+1}} - 1 \right) = x$, $x > 1$, and divergent when $x < 1$, the case $x = 1$ remaining undetermined.

We come now to the criteria of De Morgan and Bertrand which have been unsurpassed for practical adaptability to many series whose ratio of convergence is ultimately unity. De Morgan established the logarithmic scale of functional dimensions which makes $x^a (\lg x)^b$, b being positive, of a higher dimension than x^a and of a lower dimension than x^{a+k}, however small k may be and however great b may be. Between x^a and x^{a+k} may be found an infinite number of functions higher in dimension than the first and lower than the second. Built on this idea is De Morgan's test,* the series being $\dfrac{1}{\varphi(a)} + \dfrac{1}{\varphi(a+1)} + \dfrac{1}{\varphi(a+2)} + \ldots$ "First examine $P_0 = x \, \varphi'(x) / \varphi(x)$, when x is infinite. If, then, a_0, the limit of P_0, be > 1, the series is convergent; if < 1, divergent. But if $a_0 = 1$, find a_1, the limit of P_1, or $\lambda x \, (P_0 - a_0)$; then if $a_1 > 1$ the series is convergent, if < 1, divergent. But if $a_1 = 1$, find a_2, the limit of P_2 or $\lambda^2 x \, (P_1 - a_1)$; then if $a_2 > 1$, the series is convergent, if < 1 divergent. But if $a_2 = 1$, examine P_3, etc., etc.," $[\lambda^2 x \equiv \lg (\lg x)]$. De Morgan says that "if a function could be shown for which a_0, a_1, etc. *ad inf.* are severally $= 1$, this criterion does not determine whether the series is convergent or divergent." This statement left it doubtful whether such functions could be conceived or not. Ossian Bonnet † remarked that De Morgan's test never fails, except when the number of logarithms grows to infinity, "a case which is to some degree the point of junction of convergent series and of divergent series." Bonnet's view, if correct, would make De Morgan's test an absolute criterion for all series of positive terms, but Du Bois Reymond ‡ has shown that there is a region of convergence where the logarithmic criteria completely fail and has actually constructed a demonstrably convergent series for which the logarithmic criteria fail. A. Pringsheim § illustrates the same thing by the following comparatively simple series,

$$ S = \Sigma u_n = \overset{\infty}{\underset{1}{\Sigma_\nu}} \left\{ \frac{\varepsilon_0^{(\nu)}}{\nu^r a^{\nu^r + s}} + \frac{\varepsilon_1^{(\nu)}}{\nu^r a^{\nu^r + s}} + \ldots + \frac{\varepsilon_\mu^{(\nu)}}{\nu^r a^{\nu^r + s}} \right\}, $$

in which $a > 1$, $r > 1$, $s > 0$, $\mu =$ the largest integral number contained in $a^{\nu^r + s}$. Suppose, moreover, that either

---

* De Morgan's Calculus, p. 326.
† *Liouville's Journal*, vol. 8 (1843).
‡ *Crelle's Journal*, vol. 76.
§ *Mathematische Annalen*, vol. 35 (1889).

$\varepsilon_0^{(\nu)} = \varepsilon_1^{(\nu)} = \ldots = \varepsilon_\mu^{(\nu)} = 1$, or $1 + h_\nu = \varepsilon_0^{(\nu)} > \varepsilon_1^{(\nu)} > \ldots > \varepsilon_\mu^{(\nu)}$ $= 1$, h_ν being selected positive and so as to make the last term in each group of terms of the series larger than the first term in the next following group. This series is convergent, for

$$S < \sum_1^\infty \frac{(1 + h)(\mu + 1)}{\nu^r a^{\nu^{r+s}}} < (1 + h) \left\{ \sum_1^\infty \frac{1}{\nu^r} + \sum_1^\infty \frac{1}{\nu^r a^{\nu^{r+s}}} \right\},$$

the two series in the brace being evidently convergent. Let us now apply Bonnet's form of the logarithmic criteria, which says that a series is convergent if $L_\kappa(n) \lg_\kappa^\rho n \cdot a_{n+p}$, where $L^\kappa(n) \equiv \lg_0 n \cdot \lg_1 n \ldots \lg^\kappa n$, approaches a limit less than some positive finite quantity.* Let $u_n = \dfrac{\varepsilon_\mu^{(\nu)}}{\nu^r a^{\nu^{r+s}}}$ be the last term in any group (except the first). Then $n > \mu + 1 > a^{\nu^{r+s}}$. Hence $n \lg n \cdot u_n > a^{\nu^{r+s}} \cdot \lg a^{\nu^{r+s}} \cdot \dfrac{1}{\nu^r a^{\nu^{r+s}}}$ $i.e. > \lg a \cdot \nu^s$, which increases indefinitely with ν. Thus, even though the terms u_n never increase, we have among the values of $\lim (n \lg n \cdot u_n)$ the value ∞; all the more will $\lim \{ L_\kappa(n) \lg_\kappa^\rho n \cdot u_n \}$ have ∞ values, and the expression does *not* approach a limit less than some finite quantity. Thus the logarithmic tests *may* fail to indicate convergence even where the convergence can be easily established by other means. If in the above series we make $r = 1$, the resulting series is easily seen to be divergent. But the logarithmic expression $\lim \{ L_\kappa(n) \cdot u_n \}$, which indicates divergence if it is greater than some positive quantity, becomes here zero if the first term of a group be taken for u_n. Thus we have divergent series to which logarithmic criteria are inapplicable.

Some of De Morgan's results were reached independently by Bertrand. De Morgan's criteria were expressed in more convenient form by him and by Bonnet. His memoir † contains a discussion of various forms of criteria and a proof of the equivalence of his tests with De Morgan's, and that of Raabe as generalized by himself.

II. *General Criteria.* The treatment of the question of convergence from a still wider point of view, culminating in a regular mathematical theory, was begun by Kummer and continued by Dini, Du Bois Reymond, Kohn, and Pringsheim. The tests thus far given are called by Pringsheim *special*

---

* The inferior index here indicates the order of the functional operation, while the superior index is the power exponent, *e. g.* $\lg_2^3 (n) \equiv (\lg \lg n)^3$.

† *Liouville's Journal,* vol. 7.

criteria, because they all depend upon a comparison of the nth term of the series with special functions, a^n, n^κ, $n(\lg n)^\kappa$, etc. Kummer's article * antedates the papers of De Morgan and Bertrand. He establishes the theorem that the series Σu_p is convergent if a function $\varphi(p)$ can be found such that

$$\lim \varphi(p) \cdot u_p = 0 \text{ and } \lim \left\{ \varphi(p) \frac{u_p}{u_{p+1}} - \varphi(p+1) \right\} \text{ is not equal}$$

to zero. Raabe's test can be deduced from this. Du Bois Reymond divides all criteria into two classes : criteria of the *first kind* and criteria of the *second kind*, according as the general term u_p or the ratio of u_{p+1} and u_p is made the basis of research. Kummer's looks like a mixed criterion, but it is really of the second kind. The true significance of the second part of that criterion was at first overlooked. Bertrand says in his *Calcul différentiel* (vol. 1, p. 244) that the great indeterminateness in the mode of applying it is an advantage to those who know how to profit thereby. But Du Bois Reymond † points out that there is nothing indeterminate, that in the selection of the function φ we have practically the choice only between the quantities $L_\kappa(n)$ arising out of the criteria of the second kind, and if criteria of the second kind were discovered, affording more delicate tests than do the logarithmic ones, such new criteria would be embraced by the second part of Kummer's criterion.

$$\text{Lim} \left\{ \varphi(p) \frac{u_p}{u_{p+1}} - \varphi(p+1) \right\} \text{ is therefore a necessary general}$$

form of all criteria of the second kind which renders an investigation of the first part of Kummer's criterion, viz. $\lim u_p \cdot \varphi(p)$, superfluous. The above criterion, bereft of the first part, was invented anew, over half a century after its first publication, by Jensen‡, who was unaware of Kummer's researches.

Dini generalized Kummer's result, but his paper § (known to the writer only through the remarks upon it made by Pringsheim) remained unnoticed, and the same ground was traversed independently six years later by Du Bois Reymond, though with somewhat greater thoroughness. The investigations of both rest upon a comparison of the terms u_n of a series with the expression $\psi(n) - \psi(n+1)$, where $\psi(n)$ invariably either increases or decreases with n. Du Bois Reymond makes the following general statement, from which the useful criteria of the first kind can be deduced as special cases : If the terms u_n of any series of positive terms be brought to the form

$$\frac{1}{\lambda n} \left\{ \psi(n) - \psi(n+1) \right\},$$

---

* *Crelle*, vol. 13 (1835).   ‡ *Comptes Rendus*, vol. 106 (1888).
† *Crelle*, vol. 76 (1873).   § *Annali dell' Univ. Tosc.*, vol. 9.

then the given series converges whenever limit $\psi(n)$ remains finite, and limit λn does not become zero ; the series diverges whenever limit $\psi(n)$ becomes infinite and limit λn does not become infinite.   This formula is outwardly the same for divergence as for convergence, but in reality it differs considerably, since $\psi(n)$ invariably increases for divergence and decreases for convergence.   Criteria for continually weaker degrees of convergence or divergence are obtained by substituting for $\psi(n)$ a succession of functions continually decreasing or increasing more and more slowly.   By selecting for such a succession

$$ e^{\mp \alpha n}, \; n^{\mp \alpha}, \; (\lg n)^{\mp \alpha}, \; \ldots \; (\lg_r n)^{\mp \alpha}, $$

where α is a positive and finite number, the logarithmic criteria are obtained.   By putting $\psi(n) = u_n \varphi(n)$, where $\varphi(n)$ is positive, Du Bois Reymond's general criteria of the second kind are obtained : if

$$ \lim_{n=\infty} \left\{ \varphi(n) - \frac{u_{n+1}}{u_n} \varphi(n+1) \right\} > 0, $$

then $\Sigma \, u_n$ is convergent.   But it is divergent if this limit is negative and $\varphi(n)$ is selected so as to render $\Sigma \dfrac{1}{\varphi(n)}$ divergent.   The researches of Du Bois Reymond were continued by G. Kohn,[*] who showed how from any convergent (divergent) series a new series can be obtained which converges (diverges) more rapidly or less rapidly than the original series, according to the nature of the function introduced in the process.   He thus arrives at a new criterion.

A remarkable advance in the general theory of convergence and divergence was made by A. Pringsheim [†] in an article of 100 pages which establishes a simple, more coherent, and more complete general theory.   He criticises Du Bois Reymond's theory because the convergent criteria in it are heterogeneous in nature to the divergent criteria ; because it does not disclose the existence of general disjunctive criteria in which the decision as to convergence or divergence can be reached from the examination of one and the same expression ; because the general criteria of the second kind do not flow naturally from those of the first kind.   Pringsheim endeavors to steer clear of these objections in his own theory, of which what follows is a very meagre outline : Let $\Sigma \, a_\nu$ be a series of positive terms, then the simplest types of criteria of the first kind may be

---

[*] *Grunert's Archiv*, vol. 67 (1882).
[†] *Mathematische Annalen*, vol. 35 (1889).

expressed in several equivalent forms of which the following are two.

$$\Sigma a_\nu \begin{cases} \text{diverges if } a_{\nu+p} \geq h \cdot d_\nu, \text{ or if } \lim \dfrac{a_{n+p}}{d_n} \geq g > h \\[2ex] \text{converges if } a_{\nu+p} \leq H \cdot c_\nu, \text{ or if } \lim \dfrac{a_{n+p}}{c_n} \leq G < H \end{cases}$$

where $n = \infty$, h and H are finite positive quantities, $d_\nu = D_\nu^{-1}$ is the general term of a divergent series, $c_\nu = C_\nu^{-1}$ is the general term of a convergent series and p represents any constant positive integer. After proving that the general term of any divergent series Σa_ν, can be expressed in the form $\dfrac{M_{\nu+1} - M_\nu}{M_\nu}$, and that the term of any convergent series Σa_ν can be expressed in the form $\dfrac{M_{\nu+1} - M_\nu}{M_{\nu+1} \cdot M_\nu}$, and where M_ν is positive and finite for finite values of ν, and invariably increases with ν, being ∞ when ν is ∞, he writes the above general criterion thus,

$$\Sigma a_\nu \begin{cases} \text{diverges if } \lim \dfrac{M_n}{M_{n+1} - M_n} \cdot a_{n+p} \geq g > 0 \\[2ex] \text{converges if } \lim \dfrac{M_{n+1} M_n}{M_{n+1} - M_n} \cdot a_{n+p} \leq G < \infty \end{cases}$$

Putting $M_\nu = \nu$, and replacing ν successively by $\lg \nu$, $\lg_2 \nu, \ldots \lg^\kappa \nu$, he deduces Bertrand's and Bonnet's forms of logarithmic criteria and then reaches a remarkable generalization constituting a new general criterion of the first kind, analogous to Kummer's criterion of the second kind, viz., the series Σa_ν is always convergent if a positive number $\varphi(\nu)$ exists for which

$$\lim \frac{\lg \dfrac{1}{\varphi(n) a_{n+r}}}{\overset{n}{\underset{0}{\Sigma_\nu}} \dfrac{1}{\varphi(\nu)}} > 0$$

The other special criteria are also deduced from the general form. The general type of criteria of the second kind is

$$\Sigma a_\nu \begin{cases} \text{diverges if } \lim P_n (D_n a_{n+p} - D_{n+1} a_{n+p+1}) < 0 \\ \text{converges if } \lim P_n (C_n a_{n+p} - C_{n+1} a_{n+p+1}) > 0 \end{cases}$$

where P_n may be any positive factor. If we take

$P_n = \dfrac{1}{a_{n+p+1}}$, the criteria become

$$\Sigma\, a_v \begin{cases} \text{diverges if } \lim \left(D_n\, \dfrac{a_{n+p}}{a_{n+p+1}} - D_{n+1} \right) < 0 \\[2ex] \text{converges if } \lim \left(C_n\, \dfrac{a_{n+p}}{a_{n+p+1}} - C_{n+1} \right) > 0 \end{cases}$$

From this Kummer's and the special criteria of the second kind are deduced, including those of Gauss.

As regards the scope of criteria of the first and of the second kind, we find that the first will always be decisive whenever the second are, but the second may fail ever so often when the first do not. Though more limited, the criteria of the second kind are nevertheless of value, for they often yield results more easily and quickly. This narrower range of application is due to the fact first clearly pointed out by Pringsheim that the terms a_v of a series may lie always above or always below the corresponding terms b_v of another series and yet there may be no fixed relation whatever between the ratios $\dfrac{a_{v+1}}{a_v}$ and $\dfrac{b_{v+1}}{b_v}$. Since in a series of positive terms the order in which the terms come has nothing to do with the convergence or divergence of the series, it is clear that $\dfrac{a_{n+1}}{a_n}$ does not usually approach a limit. Thus the case $\dfrac{a_{v+1}}{a_v} >$ or $< \dfrac{b_{v+1}}{b_v}$ is only a very special one, and the probability that the series $\Sigma\, a_v$, yielding to the tests of the first kind, yield also to those of the second kind, is very small.

The range of special tests of the first kind has been partly considered in connection with logarithmic criteria. Pringsheim points out that the logarithmic criteria have been much overvalued, that they are applicable only to series whose terms are *essentially* decreasing, so that the increase or decrease in the values of terms does not exceed certain limits. The terms must be always above or always below the corresponding terms in $\Sigma\dfrac{g}{L_\kappa(v)}$ or $\Sigma\dfrac{G}{L_\kappa(v)\, \lg_\kappa^\rho(v)}$. But this property depends again upon the order of the terms. Suppose a certain order is favorable for the use of these tests, then a promiscuous displacement of terms may render the logarithmic scale or any other scale wholly inapplicable, and all this without altering the convergence or divergence, or even the sum of the series. This failure is in no way due to the *form* of the *general* criteria, for Pringsheim shows that for every $\Sigma\, a_\nu$, *practical* criteria of the first and second kind do exist ; but his proof of this fact yields no method of finding them, except when he knows beforehand the very thing to be determined, namely, whether $\Sigma\, a_\nu$ be convergent or not !

In addition to the criteria of the first kind and second kind, Pringsheim establishes an entirely new criterion of a *third kind* and also *generalized criteria of the second kind*, which apply, however, only to series with never increasing terms. Those of the third kind rest mainly upon the consideration of the limit of the difference, either of consecutive terms or of their reciprocals. In the generalized criteria of the second kind he does not consider the ratio $\dfrac{a_{\nu+1}}{a_\nu}$ of two consecutive terms, but the ratio of any two terms, however far apart, and deduces, among others, two criteria previously given by Kohn * and Ermakoff,† respectively.

COLORADO COLLEGE, *April* 12, 1892.

---

* *Grunert's Archiv*, vol. 67, pp. 63–95.
† *Darboux's Bulletin*, vol. 2, p. 250 ; vol. 17, p. 142.

THE INTERNATIONAL CONGRESS OF MATHE-MATICIANS IN PARIS.

AT the Zürich Congress of 1897 it was agreed to hold the next congress in Paris in 1900, the French Mathematical Society being charged with the preparations. Circulars have been issued at intervals during the last eighteen months, calling the attention of mathematicians to the arrangements in progress. The congress was finally announced for August 6th–11th, and the opening general meeting was held in the Palais des Congrès, in the Exhibition grounds, at 9.30 on the morning of Monday, August 6th. M. Poincaré was elected President, M. Hermite, who of course was not present, being the *Président d'honneur*. The executive board was constituted as follows: vice-presidents, MM. Czuber (Vienna), Geiser (Zürich), Gordan (Erlangen), Greenhill (London), Lindelöf (Helsingfors), Lindemann (Munich), Mittag-Leffler (Stockholm), Moore (Chicago, absent), Tikhomandritzky (Kharkoff), Volterra (Turin), Zeuthen (Copenhagen); secretaries, MM. Bendixson (Stockholm), Capelli (Naples), Minkowski (Zürich), Ptaszycki (St. Petersburg), Whitehead (Cambridge, absent); general secretary, M. Duporcq (Paris). After the announcement of the officers of the sections and the names of the official delegates, and a very few words from the President, the two addresses of the day, both in French, were delivered by MM. M. Cantor (Heidelberg) and Volterra (Turin); each occupied about three-quarters of an hour.

M. CANTOR: *Sur l'historiographie des mathématiques.*

During the century drawing to its close the character of mathematics has changed; its devotees are now differentiated into geometers, analysts, algebraists, arithmeticians, astronomers, theoretical physicists, and historiographers. These last make no claim to advancing the science itself; they press neither towards the arctic pole of the theory of functions, nor towards the antarctic pole of algebra; they explore neither the steep surfaces of geometry nor the depths of differential equations. Their task is rather to draw up guides and charts, to indicate by what routes the results have been obtained, and what important points have been passed by without sufficient exploration. This work began with the History of Eudemus of Rhodes, B. C.

Reprinted from Bull. Amer. Math. Soc. **7** (November 1900), 57–79

300, of which only a fragment has been preserved, just sufficient to excite lively regret for the loss of the whole. During the next two thousand years there were many bald chronicles of mathematics, but historiography as a science begins with Montucla. Notwithstanding the errors, unavoidable at that time, to be found in the two volumes of his Histoire des mathématiques (1st edition, 1758, 2d edition with two volumes by Lalande, 1799), Montucla "est encore et restera peut-être toujours un modèle que tout historiographe des sciences doit suivre." Kästner published four volumes of his Geschichte der Mathematik in the last four years of his life, 1796–1800. He has been alternately over-praised and depreciated ; Gauss referred to him as the best poet among the mathematicians, the best mathematician among the poets of his day. His history is no real history, it is rather a catalogue *raisonné*, but it is nevertheless invaluable, on account of its conscientious analysis of a number of works, which, with their authors, would be otherwise absolutely unknown to us now. At about the same date, 1797–1799, there appeared the two volumes of Cossali's Storia critica dell'algebra, dealing exhaustively with the period 1200–1600 ; as regards Italy only, it is true, but then during this period the Italian algebra was of importance far surpassing that of any other country. Cossali's labors for the elucidation of Leonard of Pisa and Cardan are of special merit.

Bossut published in 1810 his Histoire des mathématiques ; in this he gives only rapid *aperçus* of the general development, interesting to those that know already, useless to those that need to learn. In the present century we have first Chasles, to whom the speaker paid a warm personal tribute. In his Aperçu historique published in 1837, the notes, dealing with geometry, calculation, algebra, mechanics, which attain the dimensions of memoirs, form the model part of the volume, the text, the actual "Aperçu," being but a very condensed statement of the history of synthetic geometry. The other historical work of Chasles, the Rapport sur les progrès de la géométrie of 1870, is seriously affected by his ignorance of the German language. The years 1837–1841 saw the publication of Libri's Histoire des sciences mathématiques en Italie, from the earliest times up to the middle of the 17th century, a work which owing to the author's admirable style "se lit comme un roman, même dans les parties où elle n'en est pas un." Notwithstanding Libri's immense services in the study of manuscripts, his history is vitiated, as a historical

work, by his misplaced patriotism ; according to him all
progress in mathematics is due to the Italians, with perhaps
a few scattered French writers.   When he finds an Italian in
possession of any ideas or methods, no matter whence derived, he at once credits him with their discovery.   In any
case, it is not possible to give any true idea of the history of
mathematics by tracing it in one country only.   If there
is an international science, it is mathematics ; it bears
no stamp of nationality.   In considering the earliest times,
it is impossible to understand the course of mathematics in
one country without following it in others also ; to understand Greek mathematics, we must know something of
Egypt and Babylonia ; the mathematics of the Arabs cannot
be explained without reference to Egypt, Greece, and India.
After the invention of printing, so long as Latin was in use,
mathematics had no country ; and even when the frontiers
were faintly marked by the use of different languages, they
were speedily obliterated for most mathematicians.

Passing rapidly over Gerhardt and Quetelet, with a few
words of recognition, M. Cantor spoke of Nesselmann's
Die Algebra der Griechen, 1842, "un chef d'œuvre digne
d'être mis a côté de l'Aperçu historique de Chasles "; of
Arneth's Geschichte der reinen Mathematik, 1852, which
would have been an excellent book, if the author had made a
better apportionment of his space to his material—parts of the
work "fourmillent de remarques aussi spirituelles que pro-
fondes "; of Hankel's posthumous fragment, 1876, "un
torse d'une telle beauté qu'il eut été pitié de ne pas le met-
tre au grand jour "; and of the prince Baldassare Boncom-
pagni's disinterested labors on behalf of historiography.
In this sketch he passed over many authors "tous aussi
morts que leurs livres ; gardons-nous de les ressusciter ";
and avoided all mention of living authors for very obvious
reasons.   He brought his address to a close by a forecast of
the mode in which the history of more recent mathematics
must be written.   Regarding Lagrange as the founder of
modern mathematics, this gives 1759 as the starting point ;
and from this year on, the different subjects will have to be
treated in special volumes.   This however will be insufficient ; the development of the lines of thought that run
through all these different branches of mathematics must
be traced in one final volume, the History of Ideas ; difficult
to write, certainly, but indispensable, for as Jacobi said,
" Mathematics is a science of which it is impossible to understand any one part without knowing all the others."

V. Volterra: *Trois analystes italiens; Betti, Brioschi, Casorati.*

The scientific existence of Italy as a nation dates from a journey which Betti, Brioschi, and Casorati took together in the autumn of 1858, with the object of entering into relations with the foremost mathematicians of France and Germany. It is to the teaching, labors, and devotion of these three, to their influence in the organization of advanced studies, to the friendly scientific relations that they instituted between Italy and foreign countries, that the existence of a school of analysts in Italy is due.

The extent of their joint influence, affecting minds of many diverse casts, is largely due to the differences in their natural faculties, in the circumstances of their lives, and in their acquired tendencies. Brioschi, "toujours jeune par son caractère et toujours mûr par son esprit," a Lombard by birth, was at first an engineer; but at an early age he acquired a profound knowledge of the classical mathematical works, and was called to the chair of mechanics at Pavia at the age of 25. He founded the Polytechnic School at Milan, and held the directorship until his death; in his capacity of Senator, he was active in public affairs; he found time to engage in public works and in engineering; and up to the last, as Director of the *Annali di Matematica*, and President of the Accademia dei Lincei, he was one of the leaders of the mathematical movement in Italy. A great contrast to this active life is offered by the calm existence of Betti. He was born in a mountain village in Tuscany; at 34 he became a professor in the University of Pisa, and at 41 Director of the Scuola normale superiore of Pisa, whose organization is much like that of the École normale supérieure of Paris; he took no part in political movements. He loved scientific researches for their own sake exclusively, without regard to the results they might produce in the scientific world, or to their importance in teaching. He did not care for publishing his researches; and even when he did undertake this, he was apt to push it aside, attracted by new ideas. The knowledge that his intellectual conception could be realized was all-sufficient for him; he did not give himself the trouble of carrying it out in detail. When once he had obtained a clear vision of hidden truths, and had constructed in his own mind a system in which they proceeded directly from the simplest principles, "tout était fait pour Betti."

Casorati was born and lived at Pavia; he passed through the various grades in the University, where at the time of his death he was professor of infinitesimal analysis.   He lived and worked almost exclusively for his pupils ; all his works bear the stamp of the practical teacher, bent on elucidating some obscurity, correcting some error, expounding some theory.   All his writings were in a definite relation to his university teaching ; in his mind there was no distinction between the work of the savant and the work of the professor.

The fundamental differences in the three can be brought out most clearly by a comparison of their attitude towards the theory of functions.   The development of this theory exhibits three well-marked periods, corresponding to the three distinct phases that can be recognized in the history of any mathematical subject ; these three phases, however, correspond also to three distinct modes of regarding questions in analysis, each of which has its advocates.   In the first instance, the discovery of facts is all-important, and particular theories are elaborated.   There are no uniform methods ; every question is attacked on its own merits. and methods are created as occasion arises ; the ideas and results disengage themselves finally from long calculations.   In the theory of functions this manifests itself in the heroic period, personified in Euler, Jacobi, Abel ; and this manner of approaching questions is natural to Brioschi, the engineer and practical man, with his extraordinary gift for dealing with formidable calculations.   He remained faithful to the classical method, never attracted by the second phase, which he even scorned somewhat.   In this second phase, ideas replace calculations ; the philosophic spirit takes control and demands a general method including the whole subject in one body of doctrine.   This desire found its fulfilment in the second period of the theory of functions. in the works of Cauchy, Weierstrass, and Riemann, who derive everything from the very sources of the fundamental conceptions.   To this period belongs Betti the philosopher.   His broad and cultivated mind loved philosophic systems ; his Tuscan indolence (which is not intellectual idleness) caused him to delight in meditation rather than in mechanical labor.   Curiously enough, his name is associated with the theory of Weierstrass just as surely as with that of Riemann ; his education had made him an algebraist while nature meant him for a physicist.

In the final period the theories find their appropriate applications, their most suitable forms ; they are refined by criticism, and cast into a didactic mould.   The name of Casorati, critic and teacher, is associated with this third

phase. His work, Teorica delle funzioni di variabili com-
plesse, has served more than any other one book to popularize
in Italy the fundamental conceptions of the theory of functions, for the reason that, while reading it, difficulties disappear. The influence of this book is not confined to pro-
fessed analysts; anyone attempting to trace the development of mathematics in Italy during this half century will
find that analysts and pure geometers have influenced one
another. For instance, the ideas of Riemann are at the
foundation of many of the works of Italian geometers, and
while the actual introduction of these ideas was due to
Betti, it is this book of Casorati's that has carried them
everywhere and attracted the attention of geometers.

This comparison of the work of these three analysts in
the region that they had in common gives no idea, however,
of the extent of the labors and influence of each one. For
this it would be necessary to dwell on the work of Casorati
in the theory of differential equations, in analytical and
infinitesimal geometry; of Betti in mathematical physics
and algebra, he being one of the first to accept the new
ideas of Galois; of Brioschi in mechanics, algebra, and geometry. The field in which Betti and Brioschi first obtained
renown was in fact that of algebra; their names will always
be associated with that of Kronecker as second only to Hermite in their work on the equation of the fifth degree, an
equation whose complete solution was due to and secured
immortality for M. Hermite.

This concluded the business of the first general meeting,
with the exception of one or two formal announcements relating to secretarial matters. This was the only one of the
meetings to be held in the Exhibition grounds; all the
others were held at the Sorbonne. Six sections were organ-
ized for the presentation of special papers, to meet on the
7th, 8th, 9th, and 10th of August, as follows:

Section I, Arithmetic and Algebra; Tuesday, Thursday
and Friday mornings; president, M. Hilbert, secretary,
M. Cartan.

Section II, Analysis; Tuesday and Thursday mornings;
president, M. Painlevé, secretary, M. Hadamard.

Section III, Geometry; Tuesday and Thursday afternoons; president, M. Darboux, secretary, M. Niewen-
glowski.

Section IV, Mechanics and Mathematical Physics; Tuesday and Thursday afternoons; president, M. Larmor, secretary, M. Levi-Cività.

Section V, Bibliography and History; Wednesday morning and afternoon and Friday morning ; president, Prince Roland Bonaparte, secretary, M. d'Ocagne.

Section VI, Teaching and Methods ; Wednesday morning and afternoon and Friday morning ; president, M. Cantor, secretary, M. Laisant.

Sections V and VI, however, amalgamated and sat as one section, thus making the sections the same as those at Zürich. It hardly seems advisable to give a complete list of the papers, as all will be given in the full official report, which will appear shortly ; it seems better to give some account of the most interesting.

In Section I the most noticeable communication was that of M. Henri Padé, of Lille, "Aperçu sur les développements récents de la théorie des fractions continues." The object of this communication was the discussion of the question as to what is to be understood by the development of a function as a continued fraction, and the examination of the consequences of the answer obtained. For the function e^x, for instance, five such developments are already known, due to Euler, Lagrange, and Gauss. Why five? By what are the five characterized? How are they related to one another? The bond that unites them consists in a certain property of approximation common to their convergents, and thus there arises the more general question of the study of rational fractions satisfying this condition of approximation for a given function.* This investigation yields the following results. Every function that is developable by Maclaurin's formula gives rise to an infinity of developments as a continued fraction ; as to form, these fractions present the common characteristics that all the partial numerators are monomials, with coefficient and exponent different from zero ; as to matter, they are characterized by the property that all the convergents satisfy the condition of approximation mentioned above, and give an approximation whose order increases constantly as we pass from any convergent to the following one.

Among these fractions, called *fractions continues holoïdes* of the function, are included the regular continued fractions, and it is among these last that are found, as a very particular case, the developments in continued fractions already known for some special functions, for example, the five developments of e^x.

* "Sur la représentation approchée d'une fonction par des fractions rationnelles," *Ann. scient. de l' École norm. sup.*, vol. 9 (1892).

M. Padé then indicated the two ways in which these results can be generalized, the extension they involve in all the applications of continued fractions hitherto made, and the important consequences to which they lead, both in the theory of functions, where they have already introduced the question of the use of divergent power-series, and in the theory of numbers.

In Section II, the first paper read on Tuesday morning was M. Tikhomandritzky's "L'évanouissement des fonctions θ de plusieurs variables indépendantes." The function

$\theta\left(u_h \overset{\xi p}{\underset{1\ x_0}{-}} I_h\right)$ of p independent variables $u_h = \sum\limits_{i=1}^{p} \overset{x^i}{\underset{a_i}{I_h}}$ vanishes

1° when some of the points $\left(\overset{p}{\underset{1}{x_i, y_i}}\right)$ fall at (ξ, y_ξ) : 2° when

they are on an adjoint curve of the first kind, $\varphi(x^{m-2}, y^{n-2}) = 0$. If with Weierstrass we define this function by the equation

$$\theta\left(u_h \overset{p\ \xi}{\underset{1\ x_0}{-}} I_h\right) = e^{\int \Sigma\left[c_k + J\left(u_h \overset{p}{\underset{1}{-}} I_h\right)_k\right] du_k} , \tag{1}$$

where
$$J\left(\overset{p}{\underset{1}{u_h}}\right)_k = \sum\limits_{i=1}^{p} \overset{x_1}{\underset{x_0}{\Pi_k}} \tag{2}$$

($\overset{x_i}{\underset{x_0}{\Pi_k}}$ denoting the integral of the second species, which becomes infinite when (x_i, y_i) falls at (a_k, b_k)), this property of θ must be derived from those of the function (2). For this purpose, considering in the first place the function

$$J\left(u_h \overset{p\ x'}{\underset{1\ \xi}{+}} I_h\right)_k = \sum\limits_{i=1}^{p} \overset{a_i}{\underset{x_0}{\Pi_k}} \tag{3}$$

$((x', y')$ denoting a point very near to $(a_k, b_k))$, where the points (x', y'), $(\overset{p}{\underset{1}{x_i, y_i}})$ are the infinities, and (ξ, y_ξ), $(\overset{p}{\underset{1}{a_i, y_{a_i}}})$ the zeros, of the principal function of (z, y_z)

$$P_{z,\xi}\left(x', y'\ ;\ \overset{p}{\underset{1}{x_i, y_i}}\right) = \frac{P_{x', x_p}\left(z, y\ ;\ \overset{p-1}{\underset{1}{x_i', y_i'}}\ ;\ \xi, y_\xi\ \middle|\ \overset{p}{\underset{1}{a_i, y_{a_i}}}\right)}{\varphi\left(\overset{m-2\ n-2}{z,\ y_z}\ ;\ x_i, y_i\ \middle|\ \overset{p-1}{\underset{1}{x_i', y_i'}}\right)} \tag{4}$$

we see, by the second form of this function, that in the two cases one of the infinities of the function in the numerator being absorbed by one of its arbitrary zeros (1° (x_p, y_p) by $(\bar{z}, y_\xi)$; 2° (x_p, y_p) by (x'_{p-1}, y'_{p-1})), the other will be absorbed by one of its non-arbitrary zeros $\left(\overset{p}{\underset{1}{a_i, y_{a_i}}}\right)$. Hence in the two cases, at the limit, for $(x', y') = (a_k, b_k)$, one of the integrals in (3) will become infinite like $\dfrac{1}{x - a_k}$ for $x = a_k$; thus θ will vanish.

More general interest was taken in M. Mittag-Leffler's papers, which followed. ''Sur fonction analytique et expression analytique.'' '' Une application de la théorie des séries n-fois infinies.'' ''Sur une extension de la série de Taylor.'' In these M. Mittag-Leffler reported on his recent researches* in the theory of functions. Let $f(a)$, $f'(a)$,

$f''(a), \cdots$ determine an element $P(x\,|\,a) = \overset{\infty}{\underset{n=0}{\sum}} \dfrac{1}{n!} f^n(a) \cdot (x - a)^n$

of an analytic function $f(x)$. This analytic expression is valid only for its circular domain of convergence; for a more extended region $P(x\,|\,a)$ must be supplemented by certain of its continuations. Let a lie within a continuous region which does not overlap itself, and let the branch of $f(x)$ derived from $P(x\,|\,a)$ by continuation throughout K be one-valued and regular; is it possible to find a *single* analytic expression for this branch $fK(x)$, where K is given its maximum extension, the formula to be based only on the primary quantities $f(a)$, $f'(a)$, $\cdots$? M. Mittag-Leffler has shown that such an expression can be built up in a comparatively simple manner. In doing this he has employed a new geometrical notion, that of the *star* (étoile). If a ray revolves about a, proceeding in each position to the nearest singular point of $f(x)$, this point lying it may be at infinity, the collection of points on the totality of such rays forms for $f(x)$ the star A; it is assumed that the lower limit of these rays is not zero. For A we have the theorem that

$$fA(x) = \overset{\infty}{\underset{m=0}{\sum}} G_m(x),$$

where $G_m(x)$ denotes a polynomial $\Sigma c_{n,\,m} f^n(a)(x - a)^n$, in

* ''Sur la représentation analytique d'une branche uniforme d'une fonction monogène;'' *Acta Math.*, vols. 23, 24 (1899, 1900).

which the coefficients $c_{n,m}$ are given initially and do not depend on a or on $f(a)$, $f'(a)$, $\cdots$.

The expression

$$G_n(x \mid a)$$

$$= \sum_{\lambda_1=0}^{n^2} \sum_{\lambda_2=0}^{n^4} \cdots \sum_{\lambda_n=0}^{n^{2n}} \frac{1}{\lambda_1! \cdots \lambda_n!} f^{(\lambda_1 + \cdots + \lambda_n)}(a) \left(\frac{x-a}{n} \right)^{\lambda_1 + \cdots + \lambda_n}$$

leads to a limiting expression $\lim_{n=\infty} G_n(x \mid a)$ with the following properties : It is uniformly convergent for every region interior to the star A, but never uniformly convergent for a region containing a vertex of A. Within A it represents the branch $fA(x)$ of $f(x)$.

It is perfectly possible that $\lim_{n=\infty} G_n(x \mid a)$ may converge outside A; the star A is not a star of convergence for $\lim_{n=\infty} G_n(x \mid a)$. M. Mittag-Leffler has shown that it is possible to replace $\lim_{n=\infty} G_n(x \mid a)$ by another expression for which A is a star of convergence. $\lim_{n=\infty} G_n(x \mid a)$ was obtained from an n-fold series in x by making the maximum values of $\lambda_1, \lambda_2, \cdots, \lambda_n$ proceed simultaneously to the limit ∞. If the passage to the limit is performed in another way, viz., by taking $\sum_{\lambda_1=0}^{\lambda_1} \sum_{\lambda_2=0}^{\lambda_2} \cdots \sum_{\lambda_n=0}^{\lambda_n}$ in place of $\sum_{\lambda_1=0}^{n^2} \sum_{\lambda_2=0}^{n^4} \cdots \sum_{\lambda_n=0}^{n^{2n}}$ and then making $\lambda_n, \lambda_{n-1}, \cdots, \lambda_1$ tend *successively*, in the order named, to infinity, the expression

$$S_n(x \mid a) = \sum_{\lambda_1=0}^{\lambda_1} \sum_{\lambda_2=0}^{\lambda_2} \cdots \sum_{\lambda_n=0}^{\lambda_n} c_{\lambda_1 \cdots \lambda_n} f^{\lambda_1 + \cdots + \lambda_n}(a) \left(\frac{x-a}{n} \right)^{\lambda_1 + \cdots + \lambda_n}$$

(where the c's are given numerical constants of which

$$c_{\lambda_1} = \frac{1}{\lambda_1!}, \quad c_{\lambda_1 \lambda_2} = \frac{1}{\lambda_1! \, \lambda_2!} \left(\frac{1}{2} \right)^{\lambda_1 + \lambda_2}, \quad c_{\lambda_1 \lambda_2 \lambda_3} = \frac{1}{\lambda_1! \, \lambda_2! \, \lambda_3!} \left(\frac{1}{3} \right)^{\lambda_1 + \lambda_2 + \lambda_3},$$

while for values of $n > 3$ they are algebraic irrationals) yields the desired result; $\lim_{\infty} S_n(x \mid a)$ has the star A as a star of convergence, and represents $fA(x)$ within A. Writing $n = 1$, the series is seen to be simply Taylor's series ; in general it is an extension of Taylor's series.

In the course of his remarks M. Mittag-Leffler referred to recent researches of M. Borel ; this led to a discussion in which MM. Borel, Hadamard, and Painlevé took part, on the nature of the connection between "analytic expression in a complex variable x" and "analytic function in x."

In Section III papers were presented by MM. Lovett, "On contact transformations between the essential elements of space"; Macfarlane, "Applications of space analysis to curvilinear coördinates"; Stringham, "Orthogonal transformations in elliptic or in hyperbolic space"; Amodeo, and others. In Section IV very few papers were read; one appointed meeting of the section was not held, and some of the papers intended for the section were presented at the joint sitting of Sections V and VI, which was transformed momentarily into a sitting of Section IV, to hear MM. Hadamard: "Relations entre les caractéristiques réelles et les caractéristiques imaginaires pour les équations différentielles à plusieurs variables indépendantes"; and Volterra: "Comment on passe de l'équation de Poisson à caractéristique imaginaire à une équation semblable à caractéristique réelle."

The communications made in Sections V and VI, while not necessarily the most valuable mathematically, were yet of the most general interest, and lend themselves best to any general report. The sitting was opened by M. Hilbert's address, in German, on the future problems of mathematics. The lines along which we may expect the development of any science which is progressing in a continuous manner can be detected by an examination of the problems to which attention is specially paid. Among these most importance is to be attached to those that are sharply defined and stand out well; such, for example, as the problem of three bodies. In the earlier stages of any science, problems present themselves naturally through experience, as is exemplified in mathematics by the duplication of the cube and the quadrature of the circle, and at a later date by the questions arising with reference to infinitesimal analysis and the theory of the potential; but as the science progresses, it is the logical faculty of the intellect that imposes on us problems such as are found in the theories of prime numbers, elliptic functions, etc.

As to our aim with regard to any problem, there must be a definite result of some kind, it cannot be laid aside until we have obtained either a satisfactory solution or a rigorous demonstration of the impossibility of a solution. The mathematical rigor that is essential in the treatment of a problem does not require complicated demonstrations; it requires only that the result be obtained by a finite number of logical steps from a finite number of hypotheses furnished by the problem itself; in seeking this rigor we may find simplicity. The proper treatment of any problem depends on 1° a complete system of axioms, by means of which the conceptions are defined, 2° a system of symbols appropriate to the conceptions with which the problem deals; thus a demonstration by means of geometrical symbols is as legitimate as an arithmetical one, provided that the axioms on which it is based are perfectly understood. The mere formulation of these axioms is in some cases itself the problem, as for instance in arithmetic and physics. Among the ten problems that M. Hilbert specified in particular as fitted to advance mathematics, No. 2 is that of finding some one system of independent compatible axioms governing and defining arithmetical conceptions, and No. 3 is the same question for the calculus of probabilities, rational mechanics, and physics. Other problems are to prove that $e^{i\pi z}$ is transcendental when z is an algebraic irrational; and that the solution of the general equation of the 7th degree cannot be obtained by means of a finite number of operations involving only two parameters. In geometry, the relative situation of the circuits that a plane curve of assigned order can possess, with the corresponding question as regards surfaces, demands investigation; in the theory of functions there is the question of the expression of two variables, connected by any analytic relation whatever, as uniform functions of a single parameter z—for Poincaré's theorem (*Bulletin de la Société mathématique de France*, volume 11 (1883)) is subject to some limitations. These are but a few of the problems that M. Hilbert mentioned, and these were a selection from a much longer list for which he referred to an article about to appear in the *Nachrichten der Kgl. Gesellschaft der Wissenschaften zu Göttingen*, 1900. In the course of a rather desultory discussion that followed the reading of this paper, the claim was made, though apparently without adequate grounds, that more had been done as regards the equation of the 7th degree (by some German writer) than the author of the paper was willing to allow. A more precise objection was taken to M. Hilbert's remarks on the axioms of arithmetic by M. Peano, who claimed that such a system as that specified as desirable has already been established by his compatriots MM. Burali-Forti, Padoa, Pieri, in memoirs referred to on pp. 3–5 of no. 1, volume 7 of the *Rivista di Matematica*.

M. Hilbert was followed by M. Fujisawa, the official delegate from Japan, who gave, in English, a very interesting account of the mathematics of the older Japanese school. It is difficult to follow the course of Japanese mathematics ; there are some two thousand manuscript volumes still to be transcribed ; in these much valuable work is mixed up with what is purely elementary and even trivial.   The difficulty of arriving at any clear idea is greatly increased by the fact that publication of results was not customary; they were preserved to a great extent only by oral transmission.   So far as the books have been deciphered and collated, one fact stands out with ever-increasing clearness, and that is that side by side with one less important school of Japanese mathematics there exists another earlier system of mathematics of a peculiar kind, which had its origin in Japan, and was developed there entirely free from any external influences.

The mathematics of the first kind, probably derived from the Chinese at a very early date, displays a noticeable lack of rigor ; for instance, $\sqrt{10}$ is accepted as the value of π. As to content ; in algebra, the solution of simple equations and the formulæ for the sum of an arithmetical or geometrical progression were known ; in geometry, the right-angled triangle with sides proportional to 3, 4, 5 was used, with some propositions regarding regular polygons ; magic squares were discussed, even so far as those containing the first 400 numbers.   Bamboo rods were used for purposes of calculation ; these were placed one above another to indicate addition, side by side to indicate multiplication, diagonally to denote subtraction.

The other part of Japanese mathematics, that indigenous to the country, is of more importance and interest.   It appears that the mathematicians of this school made use of local value in expressing numbers, invented zero for themselves, and used the circle as the symbol for zero.   They were familiar with imaginaries and complex numbers ; and were such adepts at calculation that they found the value of π correct to 49 places of decimals.   M. Fujisawa explained that the knowledge of this part of Japanese mathematics so far obtained is very fragmentary, the unexplored part offers an attractive field of research for Japanese who may care to devote themselves to it.   It is a matter of purely historical interest, as the present teaching of mathematics in Japan is in no sense founded on it; for, very wisely as he thinks, the Japanese educational authorities made an entirely fresh start, sweeping away all trace of this older educational system.

The president of the section, M. Cantor, then spoke of the difficulties he encountered, when writing his Geschichte der Mathematik, in finding out anything about the earlier Japanese mathematics. When he did finally hear of a work of reference it turned out to be written in Japanese. With reference to the earliest use of zero, he expressed the opinion that it was probably due to the Babylonians, about 1700 B. C.

Another paper of interest in these sections was that of M. Padoa (Rome) on Friday morning: " Un nouveau système irréductible de postulats pour l'algèbre." Naming the object, *entier* (integer), two undefined derivatives, *successif* and *symétrique*, are considered. The seven postulates are

1°. If a is an integer, then suc. a is an integer.

2°. If a is an integer, then sym. a. is an integer.

3°. If a is an integer, then sym. (sym. a) $= a$.

4°. If a is an integer, then sym. $\{$suc. $[$sym. (suc. $a)]\} = a$.

5°. There exists an integer x such that sym. $x = x$.

6°. There do not exist two different integers x, y, such that sym. $x = x$, and sym. $y = y$.

7°. If a class u of objects satisfies the conditions

    (i) it contains some one integer,

    (ii) if it contains an integer x it contains also suc. x,

    (iii) if it contains suc. x it contains also x,

then every integer belongs to the class u.

These postulates define an algebraic field, whose nature is at once seen to agree with that of the natural field, suc. x being interpreted as $1 + x$, and sym. x as $- x$. M. Padoa did not get beyond this definition, possibly because he had entered so minutely into the details of the proof of the independence of the seven postulates that he had exhausted his allowance of time.

A great part of the Friday morning sitting of these two sections was devoted to the discussion of a resolution, proposed by M. Leau, in favor of the adoption of some special artificial language as the vehicle for all scientific communications. Though no particular language was named in the resolution, it was made clear that " Esperanto" was the language intended. Its advocates, MM. Leau, Padoa, Boccardi, Laisant, and others, disclaimed any wish to substitute it for natural languages, but urged its adoption as the vehicle for international intercourse ; this view they upheld with great earnestness "on behalf of humanity," as M. Laisant put it. The opposite view was upheld with equal earnestness, if less vehemence, by MM. Schroeder, Vassilief, Maggi, and others, chiefly on the ground that any such language is entirely unnecessary ; as M. Maggi remarked, mathematics already has a universal language, the language of formulæ.   In the end the suggestion of M. Vassilief was adopted, that the Congress should place itself on record as opposed to any unnecessary diversity in the languages employed, that is, practically, to the use of any language for scientific purposes other than English, French, German, and Italian, though these languages were not specified in the resolution adopted.

On Saturday, August 11th, the concluding general meeting was held at 9 a. m.   The first business was to determine the time and place for the next meeting.   At Zürich, Professor Klein, on behalf of the German Mathematical Society, had expressed their great desire that the third congress should be held in Germany ; and a definite invitation to this effect was now laid before the Congress, and unanimously accepted.   The place of meeting will probably be Baden-Baden ; the date decided upon is 1904, and the time is to be either at the beginning or the end of the summer vacation.   No other business was transacted, and the two general addresses appointed for the day were then delivered by MM. Mittag-Leffler and Poincaré.   Immediately after the conclusion of the President's address, he dismissed the Congress with the words " La séance est levée, le congrès est clos."

M. Mittag-Leffler's address was entitled " Une page de la vie de Weierstrass"; in this he considered in some detail Weierstrass's attitude towards some of the mathematical ideas of his time, illustrating by copious extracts from his correspondence; unfortunately it is not possible to give any adequate account of it.   M. Poincaré's can be given more fully.

H. POINCARÉ: *Du rôle de l'intuition et de la logique en mathématiques.*

It is obvious that there are two entirely different types of mind among mathematicians, manifesting themselves in two distinct methods of treating mathematical questions. Those of the first type are dominated by logic ; those of the second are guided by intuition.   They may be called analysts and geometers, though it is not really a question of the subject with which they deal ; the analyst remains an analyst even when working at geometry, and the geometer employing himself on pure analysis is still a geometer. Nor is the distinction a mere matter of education ; a man is born a mathematician, he does not become one; and either he is born an analyst or he is born a geometer. The two types of mind are equally necessary for the progress of the science ; each has accomplished great things that would have been impossible to the other.

At first sight the ancients seem to have all been intuitionalists, but this impression disappears on closer study. Euclid, for instance, was a logician, even though every stone of his edifice is due to intuition. The natural tendencies have not changed, only their manifestation. There has been an evolution, due to the increasing recognition of the fact that intuition cannot give rigor, nor even certainty ; a proof that relies on concrete images may be very deceptive. It was soon realized that rigor cannot be expected in the demonstrations unless it is to be found in the definitions ; so long as the objects of reasoning were given simply by the bodily senses or the imagination, there was no precise idea on which reasoning could be based. Thus the efforts of the logicians were concentrated on the definitions, one result of which is that mathematics has become arithmetized.

The question arises, is this evolution ended—have we at last attained to absolute rigor, or do we deceive ourselves as our fathers did ? Philosophers tell us that it is impossible to eliminate intuition altogether from our reasonings, for no science can spring from pure logic alone. To designate this other essential, we have no name but intuition ; but this covers many different ideas. There is (1) the appeal to the bodily senses and to imagination ; (2) generalization by induction ; (3) the intuition of pure number ; on this last a veritable mathematical method is based, while from the first two no certainty can be derived. The analysis of the present day constructs its demonstrations solely from syllogisms and this intuition of pure number ; we may say that at last absolute rigor is attained.

The philosophers now object that what has been gained in rigor has been lost in actuality ; the approach toward the logical ideal has been secured by cutting the ties with reality. For the sake of the demonstration a mathematical definition is substituted for the object, and it still remains to prove that the concrete reality answers to the definition. But as this is an experimental truth, it is not the business of mathematics to establish it. It is a great step forward to have separated these two things; nevertheless there is something in the philosophic objection.   In becoming rigorous, mathematics has assumed a certain character of artificiality; if it is clear how questions can be resolved, it is no longer clear how and why they arise.   We seek for reality; but this does not reside in the separate steps of the demonstration; it must be sought rather in the something that makes for unity.   The microscopic examination of an elephant gives no idea of the animal itself; the fairy-like structure of silicious needles which is all that is left of certain sponges cannot be understood without reference to the living sponge by which this form was imposed on the silicious particles. Logic by itself cannot give the view of the whole which is indispensable alike to the inventor and to him who desires really to understand the inventor.   Logic, which alone gives certainty, is simply the instrument of demonstration; the instrument of discovery is intuition.

But analysts also are inventors; hence they cannot always be proceeding from the general to the particular, as the rules of formal logic demand, for scientific conquests are made only by generalization.   There is however a perfectly rigorous process, that of mathematical induction, by which it is possible to pass from the particular to the general.*   For the profitable use of this, to recognize the analogies whose presence makes it applicable, the analyst must have the direct feeling for the unity of an argument, for its soul and spirit; for him the most abstract entities must be living beings.   What is this but intuition?   This however does not invalidate the distinction already drawn, for it is an intuition entirely different in nature from the sensible intuition founded in imagination alone, even though psychologists may finally pronounce it also to have a sensual foundation.   It is the intuition of pure logical form, which together with the intuition of pure number makes not only demonstration, but also discovery, possible to the analyst. Thus among the analysts inventors do exist, but not many; it remains true that the most usual instrument of invention in mathematics is sensible intuition.

On the evening preceding the formal opening of the Congress an informal reunion of the members, about half of whom were present, was held at the Café Voltaire.   On Tuesday afternoon, after the rising of the sections, the members were entertained at the École normale supérieure. At noon on Sunday, August 12th, a very successful banquet

* Poincaré, "Sur la nature du raisonnement mathématique," *Revue de métaphysique et de morale*, vol. 2 (1894), pp. 371–384.

was held at the Salle de l'Athénée-Saint-Germain, when, in the absence of M. Poincaré, M. Darboux presided very pleasantly. Toasts to those present, to the hosts, to the absent, to M. Darboux, and to the next Congress, were proposed by MM. Darboux, Geiser, J. Tannery, Stephanos, and Vassilief. Invitations to receptions held by the President of the Republic and by Prince Roland Bonaparte were accepted by a number of the members.

A very large attendance had been expected, on account of the additional attractions offered by the Exhibition ; and the answers to the circulars first sent out went far to justify this anticipation, for up to last December about 1,000 mathematicians had signified their intention of being present, with 680 members of their families. The membership fee was fixed at 30 francs, with an additional 5 francs for every member of the family. As a matter of fact, the total attendance can hardly have exceeded 250 in all. There seems very little doubt that a large proportion were kept away by distaste of the crowds that were supposed to be visiting the Exhibition, and by the rumored difficulty in obtaining accommodation, a difficulty that seems to have existed mainly in the circulars of the various agencies ; but the great heat of July certainly decided many to stay away who would otherwise have been present.

The countries represented were as follows : France, 90 ; Germany, 25; United States, 17; Italy, 15; Belgium, 13; Russia, 9 ; Austria, 8 ; Switzerland, 8 ; England, 7 ; Sweden, 7; Denmark, 4; Holland, 3 ; Spain, 3 ; Roumania, 3 ; Servia, 2; Portugal, 2 ; South America, 4 ; with single representatives from Turkey, Greece, Norway, Canada, Japan, Mexico. This list is only approximate, as no revised list of members was issued. Among the members from the United States were Professors Allardice, E. W. Brown, Dickson, Ely, Hagen, Halsted, Hancock, Harkness, Keppel, Lovett, Macfarlane, Pell, Scott, Stringham, Webster.

While the four languages, English, French, German, and Italian, were admitted on equal terms, by the constitution of the congresses, the great preponderance of French was noticeable. At Zürich, this preponderance existed in friendly intercourse, but French and German were about equally used in the communications, whereas in Paris all the general addresses, and most of the sectional papers, were in French, possibly out of compliment to our hosts. Probably at Baden-Baden French and German will be used in about equal proportions as at Zürich.

The distribution of the authors of communications among

the different countries may be of interest. The four general addresses were delivered by representatives from France, Italy, Germany, Sweden ; including these, papers were read by members from France, 13 ; Italy, 9 ; United States, 6 ; Germany, 4 ; Sweden, 4 ; Austria, 2 ; England, Greece, Holland, Japan, Portugal, Russia, Servia, and Spain, one each. This list may require some slight modification, as the only programme issued needed some corrections, but it is substantially correct. In some cases two or three communications were made by one member, thus bringing up the number of sectional papers to about 50 ; some of these were presented by title only in the absence of their authors. The general meetings occupied four hours, the sectional meetings 26. About 200 members were present at each of the general meetings ; at the sectional meetings the average attendance was about 90, and as two sections were usually sitting at the same time, this accounts very fairly for the members. About 160 were present at the banquet.

The arrangements excited a good deal of criticism. The committee of organization had doubtless special difficulties to contend with, as M. Laisant, to whom the secretarial part had been assigned, was unable to undertake it owing to the pressure of other duties. The mistake was then made of entrusting a part of this responsibility to the firm of Carré and Naud, whereas in such a case personal interest and individual responsibility are indispensable for ensuring proper attention to the various details of organization. Owing to this, members arriving in Paris had very great difficulty, during the first two days, in obtaining the necessary information in time for it to be of any service. The want of a common assembly room, where members might conveniently meet one another with or without concerted arrangement, was seriously felt. The arrangements in Zürich were so admirably complete in every point, that these defects were even more conspicuous by comparison. There is no doubt that a smaller town lends itself best to such a gathering ; it is not so much that there is less division of interests, as that the members are more in evidence, and so have a better chance of *realizing* one another. The first object of an international congress of mathematicians is to enable its members to meet one another in circumstances that shall excite the mathematical faculty, and make manifestations of mathematical interest most natural, thus encouraging the exchange of ideas both between individuals and larger groups. The mathematician who is in any degree a specialist is in general rather solitary in the average college—he would have been better off in Noah's Ark, for at the worst there would have been two of a kind. For his mind's health it is well that he should occasionally be thrown with those of kindred interests; it is well too that he should be made to feel the unity of mathematics.

The general addresses are of course under one aspect the most important part of the formal proceedings, giving broad views of the whole subject, and helping the pronounced specialist to realize his affiliations with other regions. Did a congress achieve no other end than that of evoking such addresses as those of Klein, Hurwitz, Cantor, Poincaré, Volterra, at Zürich and Paris, and presenting them to mathematicians of such widely differing interests, its existence would be amply justified. Something more might be done in a similar direction as regards the sectional meetings; an address, not necessarily by the president of the section, dealing in a broad manner with the region, or some part of the region, assigned to the section, might very well be arranged for. While the general addresses enable the individual to appreciate the relation of his special subject to mathematics as a whole, these sectional addresses would assist him in the equally important and even more difficult task of gauging his own relation to his special subject. One or two of the communications offered, both at Paris and at Zürich, were of this nature; but the matter should not be left to chance, it is well worth systematically arranging. Such an address would deal sometimes with the general ideas of the subject, sometimes with what remained to be done, sometimes with what had been accomplished during the last few years.

As to the nature of the more special papers, it hardly seems advisable to restrict the present liberty, not even to the extent that is found salutary in the regular meetings of a mathematical society; encouragement is perhaps more needed. But some control over the time consumed should be exercised; not only a theoretical control, confined to a printed statement that it must not exceed twenty minutes, but a real control, exerted as a regular thing. Ten or fifteen minutes, well employed, is quite long enough for the ordinary type of special theorem, for an outline of the method of proof is sufficient in an oral communication, and another five or ten minutes ought to enable the speaker to indicate its connections within the subject. The strict limitation to a time previously determined would in most cases be beneficial to the author, obliging him to select, subordinate, and group his details.   It is tolerably certain that if the author regards all details as equally important, his auditors will regard all as equally unimportant.

One thing very forcibly impressed on the listener is that the presentation of papers is usually shockingly bad.   Presumably the reader desires to be heard and understood ; to compass these ends, instead of speaking to the audience, he reads his paper to himself in a monotone that is sometimes hurried, sometimes hesitating, and frequently bored.   He does not even take pains to pronounce his own language clearly, but slurs or exaggerates its characteristics, so that he is often both tedious and incomprehensible.   These failings are not confined to any one nationality ; on the whole the Italians, with their clear and spirited enunciation, come nearest to being free from them.   It would be invidious and impertinent to mention names ; the special sinners sit in both high and low places.   But it is perhaps pardonable to refer to M. Mittag-Leffler's presentation of his paper to Section II as showing in how admirable and engaging a style the thing can be done.   It is not given to everyone to do it with this charm ; but there is no excuse for any normally constituted human being, sufficiently versed in mathematics, failing to interest a suitable audience for a reasonable time in that which interests himself, always provided that it be of sufficient novelty either in matter or in mode of treatment to justify him in presenting it at all.

At the Zürich Congress certain matters were energetically discussed in Section V ; extensive support was then given to resolutions in favor of constituting permanent commissions charged to consider $1°$ general reports on the progress of mathematics, $2°$ matters of bibliography and terminology, $3°$ the possibility of giving some permanent character to the Congress, by means of a central bureau or otherwise. Though these resolutions were not voted upon directly, it being felt that they required more deliberate discussion, yet at the concluding general meeting the members of the Zürich bureau were appointed a commission to consider the questions that seemed of most importance, and to furnish the Mathematical Society of France with such information on these points as might be useful in preparation for the Congress of 1900.   At the joint sitting of Sections V and VI in Paris M. Dickstein asked a question on behalf of the members, namely, was not the Congress to hear anything of the deliberations of this commission ?   No satisfactory answer was forthcoming ; M. Laisant replied simply that the Mathematical Society of France had been so taken up with material preparations for the Congress that it had not been able to enter upon any of these matters. He took the opportunity, however, of directing attention to the *Annuaire des Mathématiciens*, projected by Carré and Naud, as carrying into effect one suggestion made at Zürich. The question then dropped, but it was felt that this left matters in a very unsatisfactory state. It is to be hoped that a different report will be given in 1904, that the members of the Baden congress will not simply hear papers and meet friends, but have a chance to consider these matters of international concern. Some questions of business arise in every science ; they tend to settle themselves by a kind of tentative process, a survival of the fittest, or rather by general tacit consent. This is often the best process ; any attempt at forcing an expression of general agreement may result in checking development by encouraging too early a crystallization. But there are some matters, depending on concerted action, that are ripe for decision, and that cannot profitably be settled by any one nation for itself ; matters in which for want of a general agreement labor may be wasted. Such matters may naturally be decided by an international congress, whose decisions will simply have the force of general consent. One such question is that of a classification of mathematical sciences. At least two well-known systems are in use, and there may be others. Multiplication of systems of classification, like the multiplication of universal languages, practically destroys the good of any and all ; the congress ought to pronounce in favor of some one. As to the preparation of special reports, it seems doubtful whether this will be done best by the congress at present. In course of time it may assume academic functions and responsibilities, but it will be necessary for it to prove its continuity before it can with any propriety expect to control mathematical efforts. Encouragement and recognition would seem to be its appropriate province at present in these respects. The question as to how this continuity is to be obtained is a rather serious one, and deserving of careful discussion. A central bureau with various functions was suggested when the matter was under discussion at Zürich ; but there are objections to this. If the organizers of each congress will make it a point of honor to act on the recommendations of the preceding congress, taking into consideration the resolutions passed and doing what can be done towards carrying them into effect, possibly sufficient continuity may be attained without the red tape that would coil itself about any permanent bureau.   Some questions are better left unde-cided.   International agreement is not wanted on all points ; international rivalry and emulation still have their part to play, helped by the international friendships that are pro-moted by such gatherings as these international congresses.

In conclusion, I must express my thanks to many of those whose names appear in this report for the assistance I have received from them in its preparation.

CHARLOTTE ANGAS SCOTT.

BRYN MAWR COLLEGE,
October, 1900.

A SURVEY OF THE DEVELOPMENT OF GEOMETRIC METHODS.*

ADDRESS DELIVERED BEFORE THE SECTION OF GEOMETRY OF THE INTERNATIONAL CONGRESS OF ARTS AND SCIENCE, ST. LOUIS, SEPTEMBER 24, 1904.

BY M. GASTON DARBOUX.

I.

To understand thoroughly the progress made in geometry during the century which has recently closed it is necessary to glance rapidly at the state of the mathematical sciences at the beginning of the nineteenth century. It is known that in the last period of his life, Lagrange, fatigued by those researches in analysis and mechanics which have assured to him an immortal renown, began to neglect mathematics for chemistry which according to him was becoming as easy as algebra, for physics, and for philosophic speculations. This state of mind of Lagrange is almost always found at certain times in the lives of the greatest scholars. Those new ideas which have come to them in the productive period of youth, and which they have introduced into the common domain of knowledge, have yielded to them all that can be expected of them; the man has fulfilled his task, and feels the need of turning the activities of his mind toward entirely new subjects. This need, it is necessary to recollect, began to manifest itself with especial force in Lagrange's time. For at that time the programme of the investigations opened to geometricians by the discovery of the infinitesimal calculus appeared to be nearly exhausted. A few differential equations more or less complicated to be integrated, a few chapters to be added to the integral calculus, and it seemed that the very limits of the science would be reached. Laplace was finishing the explanation of the system of the world and laying the foundation for molecular physics. New ways indeed were opening for the experimental sciences and were preparing for the astonishing development which those sciences were to receive during the century now closed. Am-

---

* Translated with the author's permission, by Professor HENRY DALLAS THOMPSON.

Reprinted from Bull. Amer. Math. Soc. 11 (July 1905), 517–543

père, Poisson, Fourier, and even Cauchy, the creator of the theory of imaginary quantities, directed their best efforts to the application of the analytic methods to mechanics and molecular physics; they seemed to believe that just outside of this new domain, which they were in such haste to scour, the limits of theoretical science were rigidly fixed.

Modern geometry — this we must claim as its title to distinction — arrived at the very end of the eighteenth century to contribute in large measure to the rejuvenation of the whole of mathematical science by offering a new and productive field for investigation and especially by proving to us, by means of brilliant successes, that general methods do not exhaust a science, but that even in the simplest subject there remains much to be done by an ingenious and inventive mind. The beautiful geometric methods of Huygens, Newton, and Clairaut had been forgotten or neglected. The brilliant ideas introduced by Desargues and Pascal had been left without development, and seemed to have fallen on a barren soil. Carnot, with his Essai sur les transversales and the Géométrie de position, and Monge especially, with the descriptive geometry which he created and his beautiful theories upon the generation of surfaces, appeared just at this time and welded together the chain which seemed broken. Thanks to these two scholars the ideas of Descartes and Fermat, the inventors of analytic geometry, took beside the infinitesimal calculus of Leibniz and Newton the place which had been supinely lost, but which should never have been abdicated. "With his geometry," says Lagrange, speaking of Monge, "this devil of a man will make himself immortal." And indeed descriptive geometry has not only enabled us to coordinate and perfect the processes of every art " where excellence and success in work and product are conditioned by precision of form," but also it has proved to be the graphic representation of a general and purely rational geometry whose fertility has been demonstrated by numerous and important investigations.   Moreover, side by side with Monge's Géométrie Descriptive we must not forget to place his other masterpiece, the Application de l'analyse à la Géométrie ; nor must we forget that to Monge is due not only the notion of lines of curvature, but also the elegant integration of the differential equation of these lines for the case of the ellipsoid, for which, it is said, Lagrange envied him. It is necessary to lay stress upon the character of the whole of Monge's work. He, the regenerator of modern geometry, pointed out from the beginning — though his successors may have forgotten it — that the alliance between geometry and analysis is useful and productive ; and that perhaps this alliance is a condition for success to them both.

II.

Many geometers were molded in the school of Monge : Hachette, Brianchon, Chappuis, Binet, Lancret, Dupin, Malus, Gaultier de Tours, Poncelet, Chasles, etc.; among these Poncelet stands in the first rank. Neglecting everything in Monge's work that belongs to cartesian analysis or concerns infinitesimal geometry, he gave his efforts exclusively to the development of the purely geometric germs contained in the researches of his illustrious predecessor. Made prisoner by the Russians in 1813 at the crossing of the Dnieper and confined at Saratoff, Poncelet employed the leisure of captivity in demonstrating those principles which he developed in the Traité des propriétés projectives des figures, published in 1822, and in his great memoirs on reciprocal polars and harmonic means, which go back nearly to the same epoch. And thus Saratoff may be considered the birthplace of modern geometry. Rewelding the chain broken after Pascal and Desargues, Poncelet introduced both homology and reciprocal polars, thus emphasizing at the very beginning of the science those productive ideas on which the evolution of the science depended during the fifty years following.

The methods of Poncelet, presented in opposition to the analytic geometry, naturally were not received with favor by the French analysts. But the importance and novelty of these methods was such that they were not slow in stirring up in many directions the most profound researches. In discovering his principles Poncelet had been alone ; but several geometers almost at the same time came on the scene to study these principles in all their phases, and to deduce the essential results implicitly contained therein.

Gergonne was brilliantly editing, at just this time, a magazine which to-day has an inestimable value for the history of geometry. The Annales de Mathématiques, published from 1810 to 1831 at Nîmes, was for more than fifteen years the one journal in the whole world devoted exclusively to mathematical research. Gergonne, who has left to us in many regards an excellent model for a director of a scientific journal, had the faults of the qualities which made him successful, in that he collaborated, often against their will, with the authors of the memoirs submitted to him, making alterations which sometimes made an author say either more or less than he intended or desired. Nevertheless, he was keenly struck with the originality and range of Poncelet's discoveries. Already some simple methods of transformation of figures were known; homology even had been employed in the plane, but without extending it to space — as Poncelet did — but especially without its power and productivity being recognized. Moreover all these were point transformations, that is to say, they made a point correspond to a point. In introducing reciprocal polars, Poncelet showed in the highest degree the genius of the inventor; for he gave the first example of a transformation in which to a point corresponds something else than a point. Any method of transformation enables us to increase the number of theorems, but that of reciprocal polars has the advantage of causing to correspond to a proposition another proposition of totally different aspect. This was something essentially new. To exhibit this advantage, Gergonne invented the plan, since so prevalent, of articles written in double column, with the correlative propositions side by side; and his was the idea of replacing the demonstrations of Poncelet, which required the intervention of a conic or surface of the second degree, by the famous principle of duality, the significance of which, at first a little vague, was adequately explained in the discussions on the subject between Gergonne, Poncelet, and Plücker.

Bobillier, Chasles, Steiner, Lamé, Sturm and many others were at this time, with Plücker and Poncelet, assiduous contributors to the *Annales de Mathématiques*. Gergonne, having become rector of the academy of Montpellier, in 1831 was compelled to discontinue the publication of his journal. But the success which it had attained, and the thirst for research to the development of which it had contributed, had commenced to bear fruit. Quetelet had just founded in Belgium the *Correspondance mathématique et physique*. As early as 1826 Crelle brought out at Berlin the first numbers of his celebrated journal, in which he published the memoirs of Abel, Jacobi, and Steiner. Soon there were to appear also a large number of separate works in which the principles of modern geometry were presented and developed in a masterly manner.

The first of these, in 1827, was Möbius's Barycentric Calculus, a truly original work, remarkable for the depth of the conceptions and the clearness and rigor of expression ; then, in 1828, Plücker's Analytisch-geometrische Entwickelungen — the second part appearing in 1831 — shortly followed by the System der analytischen Geometrie published by the same author in 1835 at Berlin. In 1832 Steiner brought out at Berlin his great work : Systematische Entwickelung der Abhängigkeit geometrischer Gestalten von einander ; and in the following year the Geometrische Constructionen ausgeführt mittelst der geraden Linie und eines festen Kreises, in which a proposition of Poncelet regarding the employment of a single circle for elementary geometric constructions is confirmed by the most elegant examples. Finally, in 1830, Chasles sent to the Brussels academy — which, happily inspired, had offered a prize for the study of the principles of modern geometry — his celebrated Aperçu historique sur l'origine et le développement des méthodes en Géométrie, followed by the Mémoire sur deux principes généraux de la Science : la dualité et l'homographie, which was not published until 1837.

Time is here lacking to give an account of each of these beautiful investigations and to comment worthily upon them. Moreover to what could such a study conduct us except to a new verification of the general laws in the development of science. When the time is ripe, when fundamental principles have been discovered and stated, nothing stops the progress of ideas ; the same discoveries, or discoveries almost equivalent, are made almost at the same instant, and in divers places. Without attempting a discussion of this kind, which moreover might appear useless or irritating, nevertheless it is of consequence for us to bring out distinctly a fundamental difference in the tendencies of the great mathematicians who about 1830 had just given to geometry an impetus before unknown.

III.

Those who, like Chasles and Steiner, devoted their whole lives to inquiries in pure geometry, opposed to analysis that which they called synthesis ; and adopting in the main rather than in detail the tendencies of Poncelet, they proposed to institute an independent theory, a rival to the cartesian analysis.

Poncelet could not content himself with the insufficient resources furnished by the method of projection ; to reach the imaginary, he was compelled to contrive that famous principle of continuity which gave rise to the long discussions between himself and Cauchy. Suitably formulated, this principle is excellent and can render great service. But Poncelet erred in refusing to present it as a simple consequence of analysis ; and, on the other hand, Cauchy was in the wrong in not being willing to acknowledge that his objections, sound no doubt when applied to certain transcendental figures, were without force in the applications made by the author of the Traité des propriétés projectives. Whatever opinion one may form on the subject of such a discussion, it at least shows in the clearest manner that the geometric system of Poncelet rested on an analytic basis ; and we know moreover, by the unfortunate publication of the Saratoff notes, that it was by the aid of cartesian analysis that the principles which serve as the base of the Traité des propriétés projectives were first established.

Somewhat later than Poncelet — who, moreover, abandoned geometry for mechanics, where his work has had a preponderating influence — Chasles, for whom a chair in higher geometry in the Faculty of Science at Paris was created in 1847, endeavored to found an entirely independent and autonomous geometric theory. This he developed in two works of the greatest importance : in the Traité de Géométrie supérieure, which dates from 1852, and in the Traité des sections coniques, unfortunately unfinished, of which the first part alone appeared in 1865.

In the preface to the first of these works the author indicates very clearly the three fundamental points which allow to the new doctrine participation in the benefits of analysis, and which seem to him to mark a step forward in the cultivation of the science. These are :

1. The introduction of the principle of signs, which simplifies both statements and proofs, and gives to Carnot's method of transversals the full range of which it is capable ;

2. The introduction of the imaginary, which takes the place of the principle of continuity, and furnishes demonstrations as general as those of analytic geometry ;

3. The simultaneous demonstration of propositions which are correlative, that is to say, those which correspond in virtue of the principle of duality.

While homography and correlation were certainly studied by Chasles in his writings, nevertheless in his exposition he systematically set aside the use of the transformations of figures, which, as he thought, could not take the place of direct demonstrations because they masked the origin and the true nature of the properties obtained by means of them.   There is some truth in this judgment, but the very progress of the science causes us to consider it as too severe.   If it often happens that, employed without discernment, transformations uselessly multiply the number of theorems, it is not therefore necessary to ignore the fact that often also they aid us in the better understanding of the nature of the very propositions to which they have been applied.   Is it not the use of Poncelet's projection which has led to the very fertile distinction between projective and metrical properties, which moreover has made us acquainted with the great importance of the anharmonic ratio, whose essential property is found as far back as Pappus, and the fundamental rôle of which began to appear only after fifteen centuries in the efforts of modern geometry?

The introduction of the principle of signs was not so new as Chasles thought when he wrote his Traité de Géométrie supérieure.   Already Möbius in his Barycentric Calculus had answered a desideratum of Carnot, and employed signs in the broadest and most precise manner, defining for the first time the sign of a segment and even that of an area.   Later he was successful in extending the use of signs to lengths which are not upon the same line and to angles which have not their vertices at the same point.   Moreover, Grassmann, whose mind was so like that of Möbius, must necessarily have used the principle of signs in the definitions which form the foundation of his very original method of studying the properties of extension.

The second characteristic which Chasles assigns to his system of geometry is the use of the imaginary.   Here his method was truly new and he was able to illustrate it by exceedingly interesting examples.   We shall always admire the beautiful theory which he has left to us, of confocal surfaces of the second degree, where all the known properties and other new ones — as various as elegant — are derived from the general principle that they are inscribed in the same developable circumscribed to the circle at infinity.   But he introduced the imaginaries only by their symmetric functions and in consequence was not able to define the anharmonic ratio of four elements when these ceased to be real in whole or in part.   If Chasles had been able to establish the notion of the anharmonic ratio for imaginary elements, a formula which he gives in the Géométrie supérieure (page 118 of the new edition) would have given him immediately that beautiful definition of an angle as the logarithm of an anharmonic ratio, which enabled Laguerre, our regretted colleague, to give the complete solution so long sought to the problem of the transformation by homography and correlation of the relations which contain both angles and segments.

As did Chasles, so also has Steiner, that great and profound mathematician, followed the path of pure geometry; but he has neglected to give us a complete exposition of the methods upon which he relied. However, they can be characterized by saying that they rested upon the introduction of those elementary geometric forms which Desargues had already considered; upon the development which he was capable of giving to Bobillier's theory of polars; and finally upon the construction of higher curves and surfaces by the aid of pencils or nets of lower orders. Even if this were not known from more recent researches, analysis would suffice to show us that the field thus covered is coextensive with that to which the method of Descartes gives easy access.

IV.

While Chasles, Steiner, and later, as we shall see, von Staudt, applied themselves to the task of constructing a rival doctrine to analysis, thus in a way setting up one altar against another, Gergonne, Bobillier, Sturm, and above all Plücker, were perfecting the cartesian geometry and developing an analytic system somewhat adequate to the discoveries of the geometricians. It is to Bobillier and to Plücker that we owe the so-called method of abridged notation. Bobillier devoted to this method some truly original pages in the later volumes of Gergonne's Annales. Plücker began to develop it in his first volume, soon followed by a series of works in which are established the foundations of methodically modern analytic geometry. To him we owe tangential coördinates, trilinear coördinates, used in homogeneous equations, and finally the employment of canonical forms, the validity of which is recognized by the so-called method of the enumeration of constants, often misleading but very productive. All these happy inspirations served to infuse new blood into cartesian analysis and to enable it to bring out the full meaning of those conceptions which the so-called synthetic geometry had not been able to master completely. Plücker, with whom in this connection it is doubtless just to join Bobillier (who died prematurely) should be considered as the true originator of those methods in modern analysis in which the use of homogeneous coördinates permits one to treat simultaneously and, so to speak, without the reader noticing it, not only the particular figure under consideration, but also all those deduced from it by homography and correlation.

V.

From this time on a brilliant period opened for geometric research of every kind. The analysts interpreted all their results and endeavored to translate these results in their constructions. The geometricians aimed at discovering in every problem some general principle, usually demonstrable only with the aid of analysis, so that from this principle might be deduced without effort a mass of particular consequences, solidly bound together, and bound also to the principle from which they were derived. Thus Otto Hesse, the brilliant disciple of Jacobi, developed in an admirable manner that method of homogeneous variables to which Plücker perhaps had not been able to give full value. At this time also Boole discovered in Bobillier's polars the first notion of a covariant; the theory of quantics was created by the works of Cayley, Sylvester, Hermite, and Brioschi. Later Aronhold, Clebsch and Gordan, and other geometers still living, furnished the final notations, established the fundamental theorem relative to the limitation of the number of independent covariant forms, and thus gave that theory its present completeness.

The theory of surfaces of the second order, built up mainly by the school of Monge, was enriched by a mass of elegant properties, established for the most part by Otto Hesse, who found later in Paul Serret a worthy rival and successor.

The properties of the polars of algebraic curves were developed by Plücker, and still more by Steiner. The study of the cubic, which dates back to a much earlier period, was rejuvenated and enriched by a mass of new material. Steiner was the first to study by the methods of pure geometry the double tangents of the quartic, and later Hesse applied the methods of algebra to this beautiful question as well as to that of the points of inflexion of the cubic.

The notion of the class of a curve, introduced by Gergonne, and the study of a paradox partly elucidated by Poncelet, relative to the respective degrees of two curves, polar reciprocals to each other, gave birth to the researches of Plücker relative to the so-called ordinary singularities of algebraic plane curves. The celebrated formulas to which Plücker was thus led were later extended by Cayley and other geometricians to algebraic curves in space, and by Cayley and Salmon to algebraic surfaces. The singularities of higher orders were in their turn taken up by the geometers; contrary to an opinion then quite prevalent, Halphen demonstrated that not every such singularity can be considered as equivalent to a definite aggregate of ordinary singularities. These researches closed for a time this difficult and important question.

Analysis and geometry — Steiner, Cayley, Salmon, and Cremona, met in the study of cubic surfaces, and conformably to the anticipation of Steiner this theory became as simple and easy as that of the surfaces of the second order.

The algebraic ruled surfaces, so important in the applications, were studied by Chasles, by Cayley whose influence and footsteps are found in all mathematical investigations, by Cremona, by Salmon, by La Gournerie, and later by Plücker in a work to which we shall return later.

The study of the general surface of the fourth order seemed still to be too difficult; but that of certain particular surfaces of this order with multiple points or lines was commenced by Plücker for the wave surface, by Steiner, Kummer, Cayley, Moutard, Laguerre, Cremona, and many other investigators. As to the theory of algebraic curves in space, enriched in the elementary parts, it received the most marked growth later through the labors of Halphen and Noether, whom it is impossible here to separate. A new theory destined to have a great future was born with the works of Chasles, Clebsch, and Cremona; it concerns the study of all algebraic curves which can be traced upon a given surface.

Homography and correlation, the two methods of transformation which were the distant origin of all preceding researches, received from them in turn unlooked-for assistance and augmentation: They are not the only transformations that make a single element correspond to a single element, as could have been shown by a particular transformation cursorily treated by Poncelet in the Traité des propriétés projectives. Plücker defined the transformation by reciprocal radii vectores or inversion, whose great importance Sir William Thomson and Liouville were not long in showing both in mathematical physics and in geometry.   A contemporary of Möbius and of Plücker, Magnus, thought he had found the most general transformation which makes one point correspond to one point, but the investigations of Cremona show us that Magnus's transformation is only the first term of a series of birational transformations which the great Italian geometer shows us how to determine methodically, at least for the figures of plane geometry. Cremona's transformations will long preserve their importance and interest even though later investigations have shown that they always reduce to a series of successive applications of Magnus's transformation.

VI.

Every one of the works just enumerated, together with others to which we shall return later, finds its origin and, at least to a certain degree, its impetus in the conceptions of modern geometry ; but it is now fitting to indicate rapidly another source of great progress in the study of geometry.   Legendre's theory of the elliptic functions, sadly neglected by French geometricians, was developed and enlarged by Abel and Jacobi.   In the hands of these great mathematicians, soon followed by Riemann and Weierstrass, the theory of abelian functions, which algebra later endeavored to handle alone, began to bring to the geometry of curves and surfaces a contribution the importance of which will continue to grow.

Even earlier, Jacobi had used the theory of elliptic functions in the demonstration of Poncelet's celebrated theorems relating to inscribed and circumscribed polygons, introducing thus a chapter which has since been enriched by a multitude of elegant results ; already also had he obtained, by methods belonging to geometry, the integration of the abelian equations.

But it was left to Clebsch to be the first to show in a long series of articles the whole importance of Abel's and Riemann's notion of the deficiency of a curve by developing numerous results and elegant solutions in which the use of abelian integrals produced such simplicity as to make this appear their true origin.   The study of the inflexional points of the cubic curve, that of the double tangents to the quartic, and in general that of the theory of osculation which has furnished interest to so many both in earlier and recent times, was linked to the beautiful problem of the division of the periods of elliptic and abelian functions.

In one of his memoirs, Clebsch had studied the rational curves or those of deficiency zero; this led him toward the end of his regretably short life to consider what one may call rational surfaces, namely those which may be simply represented upon a plane. Here there was a great field for discovery, opened already for the elementary cases by Chasles, in which Clebsch was followed by Cremona and many other scholars. This was the occasion when Cremona, generalizing his investigations in plane geometry, brought to light if not the totality of birational transformations of space, at least some of the more interesting among these transformations. The extension of the idea of deficiency to algebraic surfaces was already begun; and some later works of great value have shown that the theory of simple or multiple integrals of algebraic differentials, in the consideration of surfaces as well as curves, will offer an extended region of important applications; but it is not in a historical account of geometry that it is proper to emphasize this matter.

VII.

While thus those mixed methods whose principal applications we have just indicated were being founded, those interested in pure geometry were not inactive. Poinsot, the creator of the theory of couples, developed by a purely geometric method, " in which," as he said, " not for a single moment is the object of the investigation lost sight of," the theory of the rotation of a solid body which the researches of d'Alembert, Euler, and Lagrange seemed to have exhausted; Chasles made a valuable contribution to kinematics in his beautiful theorems on the displacement of a solid body, which have been extended since by other elegant methods to the case where the movement has various degrees of freedom. He made known also his beautiful propositions on attraction in general, which stand without disadvantage beside those of Green and Gauss. Chasles and Steiner met in the study of the attraction of the ellipsoid and thus showed once more that geometry holds a conspicuous place in the highest questions of the integral calculus.

Steiner did not disdain to occupy himself at the same time with the elementary parts of geometry. His investigations on the contact of circles and conics, on isoperimetric problems, on parallel surfaces, and on the centroid of curvature excited the admiration of everyone by their simplicity and depth.

Chasles introduced his principle of correspondence between variable elements of two systems which has given birth to so many applications; but here again analysis claims its place to study the principle in its essence, to give it its exact form, and finally to generalize it. Exactly similar is the history of the famous theory of characteristics, and of the numerous researches of de Jonquières, Chasles, Cremona, and others, which were to furnish the foundation of a new branch of the science, viz., enumerative geometry. For many years the celebrated postulate of Chasles was admitted without objection; many geometers thought they had established it in an irrefutable manner. But, as Zeuthen then said, it is extremely difficult to make sure in demonstrations of this kind that there is not still some weak place in the reasoning which the author has not perceived; and indeed, Halphen, after some fruitless attempts, finally capped all these researches by showing clearly in what cases Chasles's postulate can be admitted, and when it must be rejected.

VIII.

Such were the principal investigations which at that time reinstated synthetic geometry in its honorable position, and assured to it during the last century the place which belongs to it in mathematical research. Numerous and illustrious laborers took part in this great geometric movement, but it must be acknowledged that it had Chasles and Steiner as its leaders. Such was the brilliancy displayed by their marvelous discoveries that, at least for a time, they threw into the shade the publications of many modest geometricians, less preoccupied perhaps in finding brilliant applications which should make geometry attractive, than in fixing this science upon an absolutely solid foundation. The work of these latter has perhaps received more tardy recognition, but their influence increases each day; and it will doubtless increase still more. To pass them over in silence would be to neglect, doubtless, one of the principal factors which will play a part in future investigations. It is primarily to von Staudt that we allude here. His geometric work was presented in two publications of great merit: the Geometrie der Lage, appearing in 1847, and the Beiträge zur Geometrie der Lage, published in 1856, nearly four years after the Géométrie supérieure.

Chasles, as we have seen, was preoccupied with founding a body of doctrine independent of the cartesian analysis, and here he was not entirely successful.   We have already indicated one of the objections that can be raised against this system, viz : imaginary elements are there defined only by their symmetric functions, and this excludes them necessarily from very many researches.   On the other hand, the constant use of the anharmonic ratio, of transversals, and of involution, which requires frequent analytic transformations, gives to the Géométrie supérieure an almost exclusively metrical character which separates it notably from the methods of Poncelet.   Returning to those methods, von Staudt applied himself to founding a geometry free from every metric relation and based exclusively on relations of position.   It was in this spirit that his first work, the Geometrie der Lage of 1847, was conceived. The author there takes as starting point the harmonic properties of the complete quadrilateral and those of homologous triangles, proved solely by theorems of three-dimensional geometry wholly analogous to those of which Monge's school made such frequent use.

In this first part of his work von Staudt entirely neglected imaginary elements.   Only in the Beiträge, his second work, by means of a very original extension of Chasles's method, did he succeed in defining geometrically an isolated imaginary element and in distinguishing it from its conjugate.   This extension, while rigorous, is laborious and very abstract.   In substance it can be explained as follows :   Two conjugate imaginary points can always be considered as the double points of an involution on a real straight line ; and just as we pass from one imaginary to its conjugate by changing i into $- i$, so we may distinguish the two imaginary points by making correspond to each of them one of the two different senses of direction that can be attributed to the line.   There is something a little artificial in this; and the development of the theory raised on such a foundation is necessarily complicated.   By purely projective methods von Staudt established a complete method of reckoning with the anharmonic ratio of the most general imaginary elements.   Projective geometry, like all the rest of geometry, uses the idea of order, and order begets number ; it is not then astonishing that von Staudt was able to create this calculus of his ; and we must admire the ingenuity which he displayed in attaining this end.   Despite the efforts of the distinguished geometers who have endeavored to simplify his presentation, we fear that this part of von Staudt's geometry, as well as the very interesting geometry of the profound thinker Grassmann, cannot prevail against the analytic methods which now have acquired almost universal favor. Life is short, the geometrician knows and practices the principle of least action. Despite these fears, which should not discourage any one, it seems to us that under the first form in which it was presented by von Staudt projective geometry should become the necessary companion to descriptive geometry, that it is destined to renew this geometry in its spirit, in its methods of procedure, and in its applications. This has been already recognized in many countries, notably in Italy, where the great geometer Cremona did not disdain to write an elementary treatise on projective geometry for the schools.

IX.

In the preceding sections we have endeavored to follow through and to point out clearly the far-reaching consequences of the methods of Monge and Poncelet. In originating tangential and homogeneous coördinates Plücker seemed to have exhausted all that the methods of projections and reciprocal polars could furnish to analysis. It remained for him, toward the end of his life, to return to his earlier investigations and give to them an extension which should enlarge in unlooked-for proportions the domain of geometry.

Preceded by numerous researches upon systems of straight lines, due to Poinsot, Möbius, Chasles, Dupin, Malus, Hamilton, Kummer, Transon, and above all to Cayley (who first introduced the idea of the coördinates of a line), investigations which have their origin partly in statics and kinematics, partly in geometrical optics — though preceded by all these, Plücker's geometry of the straight line will always be regarded as the part of his work where the newest and most interesting ideas are found. The fact that Plücker was the first to create a methodic study of the straight line is in itself important, but is nothing compared to what he discovered. It is sometimes said that the principle of duality puts in evidence the fact that the plane can be considered as an element of space just as well as the point. This is so, but in adding to the plane and to the point the straight line as a possible element of space Plücker was led to realize that any curve whatever, that any surface whatever can also be considered as an element of space, and thus is brought into existence that new geometry which already has inspired a great number of investigations and which will stimulate the authors of still more in the future.   A beautiful discovery of which we shall speak later has already connected the geometry of spheres with that of straight lines, and has led to the introduction of the idea of the coördinates of a sphere.   The theory of systems of circles is already begun ; it will doubtless be developed further when it is found desirable to study the representation, due to Laguerre, of an imaginary point in space by means of an oriented circle.

But before presenting the development of these new ideas which have vitalized Monge's infinitesimal methods, it is necessary to turn back and pick up again the history of the branches of geometry so far neglected in this discussion.

X.

Thus far we have confined ourselves to the consideration of those investigations of the school of Monge which have to do with finite geometry ; but some of Monge's disciples devoted themselves primarily to the development of those new ideas in infinitesimal geometry that had been applied by their master to curves of double curvature, to lines of curvature, and to the generation of surfaces, ideas which are developed at least in part in the Application de l'Analyse à la Géométrie.   Among these men we must cite Lancret, the author of the beautiful researches on twisted curves, and above all Charles Dupin, perhaps the only one who followed in all the paths opened by Monge.

Besides other publications, we owe to Dupin two works, of which Monge himself would not have been ashamed to be the author ; the Développements de Géométrie pure, appearing in 1813, and the Applications de Géométrie et de Mécanique, which dates from 1822.   Here is found that idea of the indicatrix which, after Euler and Meunier, was to revive and refound completely the theories of curvature, of conjugate tangents, and of asymptotic lines, which have taken such an important place in recent investigations.   Nor should we forget the determination of the surface upon which all lines of curvature are circles, nor above all the memoir on triply orthogonal systems of surfaces, where may be found not only the discovery of the triple system of surfaces of the second degree but also the celebrated theorem to which the name of Dupin will always be attached.

Under the influence of these publications and of the revival of synthetic methods, the geometry of the infinitely small again assumed in all investigations the place from which Lagrange had wished to exclude it forever. It is singular that the geometric methods thus restored were about to receive the liveliest impulse from the publication of a memoir which, at least at first sight, seemed to belong to the purest analysis; we allude to Gauss's celebrated paper, "Disquisitiones generales circa superficies curvas," which was presented in 1827 to the Royal Society at Göttingen, and whose appearance can well be said to mark an important and eventful date in the history of the infinitesimal geometry.

Thanks to this powerful help and to the writings of Monge and Dupin, the infinitesimal method received in France an impetus before unknown. Frenet, Bertrand, Molins, J. A. Serret, Bouquet, Puiseux, Ossian Bonnet, and Paul Serret developed the theory of twisted curves. Liouville, Chasles, and Minding joined them in the prosecution of the methodic study of Gauss's memoir. Jacobi's integration of the differential equation of the geodesic lines of the ellipsoid also incited many investigations. At the same time there was a great development of the problems studied in Monge's Application de l'Analyse. Some of the results already partially obtained by Monge were completed in the happiest way by the determination of all the surfaces having plane or spherical lines of curvature.

At this time lived Gabriel Lamé, one of the most acute of geometers, according to the judgment of Jacobi. He, like Charles Sturm, had commenced with pure geometry and had made most interesting contributions to this science by means of a little book published in 1817 and by his memoirs in Gergonne's Annales. Employing the results of Dupin and Binet concerning the system of confocal surfaces of the second degree, and rising to the notion of curvilinear coördinates in space, he became the inventor of a whole new theory destined to receive the most varied applications in the domain of mathematical physics.

XI.

Here also, in infinitesimal geometry, are found the two tendencies which have been noticed in the geometry of finite quantities.

Some, among whom must be placed J. Bertrand and O. Bonnet, wished to build up an independent method resting directly upon the use of the infinitely small. Bertrand's great Traité de Calcul différentiel contains several chapters on the theory of curves and surfaces which to a certain extent illustrate this point of view. The exponents of the other tendency followed the ordinary analytic way, aiming only at thoroughly understanding and exhibiting the elements which are of primary importance. This is the course which Lamé followed in introducing his theory of differential parameters. This is the course which Beltrami followed in extending with great ingenuity the use of these differential invariants to the case of two independent variables, that is to say, to the study of surfaces.

At present it seems to be the custom to prefer a mixed method, the origin of which is found in Ribaucour's writings under the name of *perimorphy*. Here the rectangular axes of analytic geometry are retained, but they are rendered movable and related with the system to be studied in the manner which appears the most advantageous. Most of the objections urged against the method of coördinates are thus removed ; and the advantages of what is sometimes called intrinsic geometry are combined with those which result from the use of the ordinary analysis. Nor is this analysis abandoned ; but the complicated calculus which it almost always introduces when applied to the study of surfaces in rectilinear coördinates disappear soonest if we use the notions of invariants and covariants of differential quadratic forms, which we owe to the investigations of Lipschitz and Christoffel that were inspired by Riemann's study of non-euclidean geometry.

XII.

The results of these numerous investigations were not long in coming. The notion of geodesic curvature, which Gauss already had found but had not published, was given by Bonnet and Liouville ; the theory of surfaces with a functional relation between the radii of curvature, begun in Germany with two propositions which would have worthily occupied a place in Gauss's memoir, was enriched with a mass of propositions by Ribaucour, Halphen, S. Lie, and others. Some of these propositions have to do with the general properties of such surfaces : others have to do with particular cases where the relation between the radii of curvature takes a particularly simple form, *e. g.*, the minimal surfaces, and surfaces with constant curvature, positive or negative.

Minimal surfaces have been the object of investigations which have made their study the most attractive chapter in infinitesimal geometry. The integration of their partial differential equation forms one of the most beautiful discoveries of Monge; but, owing to the imperfection of the theory of the imaginary, that great geometrician could not get from his formulas any method of generation of these surfaces — nor even a single particular surface. We will not here enter into the detailed history which has been presented in our Leçons sur la théorie des surfaces; but it is fitting to recall the fundamental investigations of Bonnet which have given to us, in particular, the idea of the surfaces *associated* with a given surface; also Weierstrass's formulas which establish a close bond between minimal surfaces and the functions of a complex variable; also Lie's investigations by which it has been established that even Monge's formulas could now be made the basis of a profitable study of minimal surfaces. The endeavor to determine the minimal surfaces of the lowest class or degree led to the notion of double minimal surfaces, which belongs to analysis situs.

Three problems varying in importance have been studied in this theory.

The first, relative to the determination of the minimal surfaces inscribed to a given developable with a given curve of contact, was solved by those celebrated formulas which later led to many special propositions; *e. g.*, every straight line traced upon such a surface is an axis of symmetry.

The second, formulated by S. Lie, requires the determination of all algebraic minimal surfaces inscribed in an algebraic developable, the curve of contact not being given. This problem also has been completely solved.

The third and most difficult is that problem which the physicist solves empirically by plunging a closed contour into a solution of glycerine. The problem has to do with the determination of the minimal surface passing through a given contour. It is evident that the solution surpasses the resources of geometry. But thanks to the capabilities of the highest analysis, it has been solved for certain contours in the celebrated memoir of Riemann and in the profound researches which followed or were coeval with this memoir. As to the solution for a perfectly general contour, its study has been brilliantly begun and it will be continued by our successors.

After minimal surfaces, the surfaces with constant curvature naturally attracted the attention of geometers. Bonnet's ingenious theorem connected together the surface of constant mean curvature and the surface of constant total curvature. Bour announced that the partial differential equation of the surface of constant curvature could be completely integrated. This result has not been attained ; and its attainment seems hardly probable in view of the investigation in which Sophus Lie vainly tried to apply a certain general method of integrating partial differential equations to the particular equation of surfaces with constant curvature. But while it is impossible to determine all these surfaces in finite terms, at least some of them characterized by special properties have been obtained, namely those with plane or spherical lines of curvature ; and by the use of a method which has proved successful in many other problems, it has been shown that from every surface with constant curvature can be derived an infinity of other surfaces of the same kind, and this by the use of clearly defined operations which require only quadratures.

The theory of the deformation of surfaces in Gauss's sense was also greatly enriched. To Minding and Bour is due the detailed study of that special deformation of ruled surfaces which leaves the generators straight lines. While, as has just been said, the determination of surfaces applicable upon a sphere was impossible, this determination for other surfaces of the second degree has been attacked with greater success, and particularly for the paraboloid of revolution. The systematic study of the deformation of the general surfaces of the second degree is already opened ; it is a study which will doubtless give most important results in the future.

The theory of an infinitely small deformation constitutes today one of the most complete chapters of geometry. It is the first extended application of a general method which seems to have a great future.

Being given a system of differential equations or a system of partial differential equations, sufficient to determine a certain number of unknowns, we conventionally associate with that system another system of equations, which we have called the auxiliary system, and which determines the systems of solutions infinitely near to any given system of solutions. The auxiliary system being necessarily linear, its use in any investigation throws a valuable light on the properties of the proposed system and upon the possibility of its integration.

The theory of lines of curvature and of asymptotic lines has been notably extended. Not only was it possible to determine these two sets of lines for certain particular surfaces such as Lamé's tetrahedral surfaces; but also, in developing Moutard's results relative to a particular class of linear partial differential equations of the second order, it was possible to generalize what had been obtained for surfaces with plane or spherical lines of curvature by determining completely all the classes of surfaces for which the problem of the spherical representation can be solved. The correlated problem relative to the asymptotic lines was solved also by finding all the surfaces of which the infinitely small deformation can be determined in finite terms. There is here a vast field for investigation, the exploration of which has hardly been begun.

The infinitesimal study of congruences of straight lines, begun long since by Dupin, Bertrand, Hamilton, and Kummer, is only beginning to be connected with all these investigations. Ribaucour, who in this took a leading part, studied certain classes of rectilinear congruences and in particular the so-called isotropic congruences, which come in most happily in the study of minimal surfaces.

The triply orthogonal systems which Lamé had employed in mathematical physics became the object of systematic investigations. Cayley was the first to form the partial differential equation of the third order upon which the general solution of this problem has been made to depend. The system of confocal surfaces of the second order was generalized and gave birth to the theory of the general cyclides, in which theory it is possible to use at one and the same time the resources of the metrical, projective, and infinitesimal geometries. Many other orthogonal systems were presented. Among these it is fitting to mention Ribaucour's cyclic systems for which one of the three families has circles for its orthogonal trajectories, and the more general systems for which these orthogonal trajectories are simply plane curves. The systematic use of the imaginary, which must not be excluded from geometry, rendered it possible to make all these determinations depend upon the study of the finite deformation of a particular surface.

Among the methods which made possible all these results, it

is fitting to designate the systematic use of the linear partial differential equations of the second order and of systems of such equations.   The most recent investigations show that this method is destined to lead to a revision of most of the theories.

Infinitesimal geometry could not neglect the study of the two fundamental problems set by the calculus of variations.

The problem of the shortest path on a surface was the object of masterly studies by Jacobi and Ossian Bonnet.   The study of the geodesic lines was extended, and these lines were determined for new surfaces.   The theory of assemblages made it possible to follow these lines in their course on a given surface. The solution of a problem relative to the representation of one surface upon another increased greatly the interest in the discoveries of Jacobi and Liouville relative to a particular class of surfaces where the geodesic lines can be determined.   The results from this particular case lead to the consideration of a new question, namely : To find all the problems of the calculus of variation the solutions of which are furnished by curves satisfying a given differential equation.

Finally, Jacobi's methods were extended to space of three dimensions and applied to obtain the solution of a question which presents the greatest difficulties, namely : the study of the minimum properties belonging to a minimal surface passing through a given contour.

XIII.

Among the discoverers who have contributed to the development of infinitesimal geometry, Sophus Lie is distinguished by many important discoveries which place him in the first rank. He was not one of those who exhibit in youth very marked aptitudes ; when in 1865 he left the University of Christiania, he was still hesitating between philology and mathematics. The works of Plücker first caused him to realize his true vocation ; and he published in 1869 a first investigation on the interpretation of imaginaries in geometry, and from 1870 on he was in possession of the ideas which mapped out his whole career.

I had the pleasure at that time of frequently seeing and entertaining him in Paris, whither he had come with his friend Felix Klein.   A course of lectures delivered by Sylow and attended by Lie disclosed to him the great importance of the theory of substitutions ; and the two friends studied this theory in Camille Jordan's great treatise; they recognized fully the important rôle which this theory was destined to play in many branches of mathematical science where it had not yet been applied. Each of them has had the good fortune of contributing by his publications to turning mathematical research in that direction which seemed to him the best.

As early as 1870 Sophus Lie presented to the Paris Academy of Sciences an extremely interesting discovery : Nothing resembles a sphere so little as a straight line, but nevertheless Lie conceived of a peculiar transformation which made a sphere correspond to a straight line, and therefore set up a method of linking every proposition having to do with straight lines to a proposition having to do with spheres and *vice versa*. In this most striking method of transformation each property of the lines of curvature of a surface furnishes a proposition on the asymptotic lines of the transformed surface. Lie's name will always remain attached to those recondite relations which join together those two essential and fundamental elements of geometrical investigation, the straight line and the sphere. These relations he developed in a memoir filled with new ideas, which appeared in 1872.

The researches which followed Lie's brilliant entrance into mathematics fully confirmed the hopes to which it had given rise. Plücker's conception of the generation of space by straight lines, or by arbitrarily chosen curves or surfaces, opened to the theory of algebraic forms a field which was not yet explored, which Clebsch had barely started to investigate and for which he had only just begun to fix the boundaries. But in the domain of the infinitesimal geometry this conception was given its full value by Sophus Lie. The great Norwegian geometrician then found first the notion of congruences and complexes of curves and later that of contact transformations, the first germ of which he had found for the case of the plane in Plücker. The study of these transformations led him to perfect, contemporaneously with Mayer, the integration methods which Jacobi had instituted for partial differential equations of the first order; moreover it threw the most radiant light upon the most difficult and obscure parts of the theory of partial differential equations of a higher order. In particular it allowed Lie to indicate every case in which Monge's method of characteristics is fully applicable to equations of the second order in two independent variables.

In the continuation of the study of these special transformations, Lie was led to construct step by step his masterly theory
of continuous transformation groups and to render evident the
important role which the group notion plays in geometry.
Among the essential elements of his investigations it is fitting
to mention the infinitesimal transformations, the idea of which
belongs exclusively to him.   Three great works published
under his direction by able and devoted collaborators contain
the essential part of his labors together with their application
to the theory of integration, to the theory of complex units,
and to non-euclidean geometry.

XIV.

I have thus arrived by an indirect way at the non-euclidean
geometry, the study of which daily assumes increasing importance in the researches of geometricians.   If I were the only
speaker on geometry, I should take pleasure in recalling to you
everything that has been done in this subject since Euclid or
at least since Legendre, up to the present day.   Considered in
succession by the greatest geometricians of the last century, this
subject has been progressively enlarged.   It began with the
consideration of the celebrated parallel postulate, but it has
come to include all the geometric axioms.

Euclid's Elements, which have withstood the toil of so many
centuries, will at least have had the honor of calling forth, before their end, a long series of admirably connected investigations, which not only will contribute in a most efficacious
way to mathematical progress, but also will furnish to the
philosopher the most precise and firm starting points for the
study of the origin and formation of our concepts.   I am
assured in advance that my distinguished collaborator will not
forget among the problems of the present this problem, which
perhaps is the most important, and with which he has occupied
himself with so much success; and I leave to him the task of
developing it with all the fullness which it assuredly deserves.

I have just spoken of the elements of geometry.   They have
received during the last hundred years additions which it
is not right to forget.   The theory of polyhedrons was enriched by Poinsot's beautiful discoveries on star-shaped, and
by Möbius's one-sided polyhedrons.   The transformation
methods have enlarged the aspect of this study.   We can
to-day express our view of this subject by saying: That the

first book contains the theory of translation and symmetry, that the second is equivalent to the theory of rotation and translation, that the third depends upon inversion and the theory of figures similar and similarly placed.   But it must be acknowledged that it is by the aid of analysis that the elements have been enriched by their most beautiful propositions.   For it is to the highest analysis that we owe the solution of the problem of the inscribed regular polygon of 17 sides and of the analogous problems.   To it we owe the demonstrations so long sought of the impossibility of the quadrature of the circle, of the impossibility of certain geometric constructions by the aid of the straight-edge and compasses only.   Finally, it is to it that we owe the first rigorous demonstrations of the maximum and minimum properties of the sphere.   It will be the task of geometry to enter this domain where analysis has preceded it.

What will be the elements of geometry in the course of the century just beginning?   Will there be a single elementary book of geometry?   America perhaps, with its schools freed from every program and tradition, will give us the best solutions of this important and difficult question.   Von Staudt has sometimes been called " the Euclid of the nineteenth century "; I should prefer to call him " the Euclid of projective geometry "; but is this branch of geometry, no matter how interesting it may be, called on to furnish alone the foundation of the future Elements?

XV.

The time has come to close this narrative already too long, and yet there remains a mass of interesting investigations which I may be said to have neglected under compulsion.   I should have liked to speak to you of those geometries in any number of dimensions, the idea of which goes back to the beginnings of algebra, but of which the systematic study has been taken up only in the last sixty years by Cayley and Cauchy.

This kind of research has found favor in your country, and I need hardly recall to you that our president,* after showing himself the worthy continuer of Laplace and Leverrier, in a space which he considers with us as being endowed with three dimensions, has not disdained to publish in the *American Journal* considerations of great interest on the geometries of n dimensions.   A single objection already formulated by Poisson may

* Professor Simon Newcomb.

be urged against studies of this kind: namely, the absence of all real basis, of any substratum that will allow the results obtained to be presented under a visible and in a way tangible form. The extension of the methods of descriptive geometry, and above all the employment of Plücker's notions of the generation of space will do much toward depriving this objection of its weight.

I should also have wished to speak to you of the method of equipollences, the germ of which we find in Gauss's posthumous papers, of Hamilton's quaternions, of Grassmann's methods, and in general of the systems of complex units, of the analysis situs so closely connected with the theory of functions, of the so-called kinematic geometry, of the theory of abaci, of geometrography, of geometric applications in natural philosophy and in the arts. But I would fear if I enlarged too much, that some analyst — there have been such — would accuse geometry of wishing to monopolize everything.

My admiration for analysis, become so fruitful and powerful in our time, would not permit me to conceive such a thought. But if such a reproach could be formulated to-day, I think that it should be raised, not against geometry but against analysis. The circle in which mathematical studies seemed bound at the beginning of the nineteenth century has been broken on all sides. The old problems present themselves to us under an altered form. New problems arise studied by legions of workers. The number of those who cultivate pure geometry has become very limited. This is a danger against which it is of some importance to guard. Do not let us forget that while analysis has acquired means of investigation which formerly it lacked, nevertheless it owes those means largely to the concepts introduced by the geometricians. Geometry must not remain, as it were, shrouded in its own triumph. It was in the school of geometry that we have learned, and there our successors will have to learn it, never blindly to trust to too general methods, but to consider each question on its own merits, to find in the particular conditions of each problem either a direct way toward a simple solution or the means to apply in an appropriate manner those general methods which every science should collect. Thus, as Chasles says in the beginning of the Aperçu historique: "The doctrines of pure geometry often, and in many questions, give a simple and natural way to penetrate to the origin of truths, to lay bare the mysterious chain which unites them, and to make them known individually, luminously and completely." Therefore let us cultivate geometry which has its own advantages, and this without wishing to make it equal in all points to its rival. Besides, if we should be tempted to neglect it, it would soon find, as it once has done, in the applications of mathematics the means of reviving and developing itself anew. It is like the giant Antæus, who regained his strength whenever he touched his mother earth.

Chapter 3

Fifth Degree Polynomials

Two superbly talented mathematicians. Their brilliance and potential mathematical contributions tragically lost with their deaths at young ages. As James Pierpont [6] stated in 1897 during a presentation at an AMS meeting,

> "... both Galois and Abel from their earliest youth were strongly attracted toward algebraical theories; both believed for a time that they had solved the celebrated equation of fifth degree which for more than two centuries had baffled the efforts of the first mathematicians of the age; both, discovering their mistake, succeeded independently and unknown to each other in showing that a solution by radicals was impossible; both, hoping to gain the recognition of the Paris Academy of Sciences, presented epoch making memoirs, one of which was lost forever, the other published first twelve years after the death of its illustrious author; in the case of both the passionate feverish blood of genius lead them to a premature death, one at the age of twenty, the other at twenty-six. Both died leaving behind them theories whose thresholds they had hardly crossed, theories which, belonging to the first creations of this century, will keep their memories perennially green."

The story starts two, three, or maybe even four millennia ago. This is because even though the notion of an equation had yet to be developed, at least around 400 BC, and according to Neugebauer [2], maybe even in texts of 4000 years ago, the Babylonians developed an algorithmic approach which, essentially, solved for the positive roots of quadratic equations. It took another millennium for Hindu mathematician Brahmagupta (598-665) to show how to handle negative roots.

The next step took place in the early part of the sixteenth century when S. del Ferro, and then Tartaglia, found solutions to the general cubic equations.[1] Soon after, L. Ferrari discovered how to solve the quartic equation. With all of this progress, it had to have been assumed that it was only a matter of time until algebraic expressions involving radicals could be found to handle the next challenge – the quintic equation. The search was on.

But, in spite of a quarter of a millennium of efforts by some of the best mathematicians of the time, no solution was forthcoming. Maybe, just maybe, this is because no such solution exists. Just prior to the birth of the nineteenth century, Ruffini published a two-volume treatise asserting, among other conclusions, that this is the case; a general solution involving radicals does not exist. While Cauchy and other eminent scholars of the time accepted Ruffini's argument, the skeptics were correct; in spite of his important contributions, Ruffini's proof has a serious gap.

An incorrect paper written by an insightful mathematician, however, can be more valuable than ten correct papers by others. This most certainty was true with Ruffini; his assertion changed the direction of the chase from finding a solution to showing that the quintic cannot, in general, be solved with radicals. A quarter century later the problem finally was resolved when Niels Hendrik Abel published, at his own expense, a proof supporting this negative assertion. It would be six more years until Galois would submit, only to have rejected, his spectacular paper on the theory of equations.

The story and even the reasons for the deaths of Abel and Galois, complete with the (incorrect) legend about details being dashed down just before "the duel," are reasonably well known. Among many sources is Stubhaug's biography of Abel [9] and Infeld's of Galois [1]. Another one describing the mathematics is S. Rosen's [8] paper which uses modern terminology to explain the gap in Ruffini's argument and to describe the essence of Abel's proof that, by being done before Galois', did not rely on Galois theory. Of course, a standard contemporary course in algebra presents Galois theory. Yet, it is refreshing to read expositions written in 1890s, a time much closer to the events, to find what people at that time thought about these men and their contributions. But to tell the story, Pierpont's remarks,

> "It is well known that Galois, like Abel and Ruffini, received inspiration from the writings of Lagrange regarding the algebraic solution of equations ... "

---

[1] The material in this and the next several paragraphs comes from the citation I wrote, which then was approved by the prize committee, for S. Rosen's 1999 Chauvenet Award from the MAA for his paper [8].

require including Lagrange's contributions, and they are.

The three *Bulletin* papers from the 1890s reproduced in this section, starting with "Lagrange's Place in the Theory of Substitutions" [4], followed by "On the Ruffini-Abelian Theorem" [5], and concluding with "Early History of Galois Theory of Equations" [6] describe the story about the fifth degree equations and the people involved. These papers most surely influenced the early attitudes of American mathematicians in their understanding of these events. All three are by J. Pierpont; the author of the first reproduced article in this book.

James Pierpont was born in 1858, one year after a different "James Pierpont" wrote a song you most surely have sung; "Jingle Bells." Our Pierpont, the mathematician, was a graduate of Worcester Polytechnic Institute and after receiving his Ph.D. at the Universität Wien, he spent his career at Yale University. He was active both in AMS activities and in encouraging and promoting the growing American mathematical research community. Indeed, the significant role Pierpont played in the formation of American mathematics is manifested by how often his name appears throughout this book; he was a frequent contributor to the *Bulletin*. Along with Maxime Bôcher, in 1896 Pierpont was the first AMS Colloquium Speaker.

One often finds attributed to him the quote

> *"The notion of infinity is our greatest friend; it is also the greatest enemy of our peace of mind."*

This attribution is reasonable as the phrase seems to be consistent with what an early and important proponent of the Lebesgue integral in the United States and author of a two volume series on the "Theory of functions of real variables" (vol. 1 in 1905 and vol. 2 in 1912) might say.

Bibliography

[1] Infeld, L. *Whom the Gods Love: The Story of Evariste Galois*, National Council of Teachers of Mathematics, 1948

[2] Neugebauer, O., *The Exact Sciences in Antiquity*, 2nd ed., Brown University Press, 1957.

[3] Pierpont, J., On an Undemonstrated Theorem of the Disquisitiones Arithmeticae, *BAMS*, **2** (1895), 77-83.

[4] Pierpont, J., Lagrange's place in the theory of substitutions, *BAMS* **2** (May, 1895), 196-204.

[5] Pierpont, J., On the Ruffini-Abelian Theorem, *BAMS* **3** (April 1896), 200-221.

[6] Pierpont, J., Early History of Galois Theory of Equations, *BAMS* **5** (April 1898), 332-340.

[7] Pierpont, J., Mathematical rigor, past and present. *BAMS*, **34** (1928), 23-53.

[8] Rosen, M., Niels Hendrik Abel and the Equations of the Fifth degree, *American Mathematical Monthly* **102** (1995), 495-505/

[9] Stubhaug, A., *Niels Henrik Abel and his times: Called too soon by flames afar*, Springer, 1996.

LAGRANGE'S PLACE IN THE THEORY OF SUBSTITUTIONS. †

BY DR. JAMES PIERPONT.

IN the present brief note I cannot vindicate Lagrange's right to the title of creator of the theory of substitutions; but I hope, by presenting a few examples of his methods, to show the importance of considering him from this point of view. Lagrange was led to the study of this theory by his attempts to solve equations of degree higher than the fourth. Speaking of the inherent difficulties which this thorny subject offered to the investigator, he remarks: ‡

"The theory of equations is of all parts of analysis the one, we would think, which ought to have acquired the greatest degree of perfection, by reason both of its importance and of the rapidity of the progress that its first inventors made; but although the mathematicians of later days have not ceased to apply themselves, there remains much in order that their efforts may meet with the success that one could desire. In regard to the resolution of literal equations one has hardly advanced further than one was in Cardan's time, who was the first to publish the resolution of equations of the third and fourth degree. The first successes of the Italian analysts in this branch seem to have marked the limit of possible discoveries: at least it is certain that all attempts that have been made up to the present to push back the limits of this branch of algebra have hardly served for other purposes than

---

† Read before the Yale Mathematical Club.

‡ Lagrange: Nouveaux Mémoires, Acad. Sciences Berlin, years 1770–71. Also, Œuvres, vol. III, pp. 205–421, Réflexions sur la résolution algébrique des équations.

Reprinted from Bull. Amer. Math. Soc. 1 (May 1895), 196–204

to find new methods to solve the equations of third and fourth degree, none of which seem applicable to equations of higher degrees."

In his great paper published in the Mémoires of the Academy of Sciences at Berlin in the years 1770–71, under the title " Réflexions sur la résolution algébrique des equations," Lagrange proposed to examine the different methods which one had found up to then to solve algebraic equations, to reduce them to general principles, and to show *à priori* why these methods succeeded in case of the cubic and biquadratic, but failed for equations of higher degree. To do this Lagrange took up successively the various methods proposed by Cardan, Ferrari, Descartes, Tschirnhaus, Euler, and Bézout, and showed that the roots of the various resolvents upon whose solution the solution of the given equation depended were *rational* functions of the roots of the given equation. Here, then, was a great and fundamental step in advance. The problem of the solution of equations was shown to depend upon the properties of *rational* functions of the roots. To study the properties of these functions, Lagrange invented a " *calcul des combinaisons*," as he styled it, which was nothing else than the first rudiments of the theory of substitutions.

By means of this new *calcul* Lagrange was placed in a position to tell in advance the result and character of certain investigations, in much the same way that algebra serves for numerical problems. Lagrange himself characterizes his method in the following words:

" These, then, if I mistake not, are the true principles for the resolution of equations. The analysis is reduced, as is seen, to a species of calculus of combinations by means of which one finds *à priori* the results one should expect." Lagrange gives the new calculus a broad and solid basis. Among the various theorems he established for rational functions of the roots of a general equation of nth degree, one is of fundamental importance: a function V which takes on n ! values for the substitutions of the symmetric group is root of an irreducible equation of degree n ! whose coefficients are rationally known. The roots of this equation are rational functions of one another, and possess the remarkable property that every rational function of the roots of the given equation can be expressed rationally in one of them, and hence the roots themselves. Similar functions he defines as those having the same group. Two similar functions are rationally expressible by each other. A rational function of the roots which takes on ρ values for the symmetric group, is root of an equation of degree ρ, whose coefficients are rational in the coefficients of the given equation. Further, ρ is a divisor of n !

If ϕ and ψ take on respectively $m\rho$ and ρ values for the symmetric group, then ϕ is a root of an equation of mth degree whose coefficients are rational in ψ. The function ψ can be expressed rationally in ϕ.

Let us see how such theorems as these enable Lagrange to assign *à priori* the reason for the success of the various methods proposed to solve the cubic and biquadratic, and their failure when applied to equations of higher degrees. I have already remarked that Lagrange found that the resolving functions employed by his predecessors were rational. In final analysis, he found that they all belonged to the type

$$t = x_1 + \alpha x_2 + \alpha^2 x_3 + \ldots + \alpha^{n-1}x_n,$$

where $x_1, x_2, \ldots x_{n-1}$ are the roots of the given equation $f(x) = 0$ and α is an imaginary nth root of unity. Let us see to what equations these functions lead. Two cases present themselves according as n is prime or composite. Let first n be prime. Then the function

$$\theta_1 = t_1^n = (x_1 + \alpha x_2 + \ldots + \alpha^{n-1}x_n)^n$$

remains unchanged for the cyclic substitutions

$$| z, z + a | \quad (\text{mod } n) \quad a = 0, \quad 1 \ldots n - 1.$$

For the substitutions

$$| z, bz | \quad (\text{mod } n) \quad b = 1, 2 \ldots n - 1,$$

θ_1 takes on $n - 1$ values

$$\theta_1 \theta_2 \ldots \theta_{n-1},$$

obtained by replacing α by respectively α^2, α^3, $\ldots$
Consider the equation

$$\theta = (\theta - \theta_1) \ldots (\theta - \theta_{n-1}) = 0.$$

Its coefficients are symmetrical functions of $\theta_1, \theta_2, \ldots$. Let ψ be such a function. It is root of an equation of degree $\rho = (n - 2)!$ If ψ can be found in any way, the coefficients of $\theta = 0$ being rational in ψ, are rationally known, and the solution of $f(x) = 0$ depends now upon an equation of degree $n - 1$.

When $n = 3$, $\rho = 1$; $n = 5$, $\rho = 6$.

Thus for the cubic we see that the coefficients of $\theta = 0$ are rational, and the solution depends therefore upon an equation of the second degree only. As soon, however, as the prime

$n > 3$, the coefficients depend upon an equation of degree higher than the given equation.   For $n = 5$ it is already of sixth degree.   In passing I note that Lagrange by the considerations of his new *calcul* made the solution of the quintic depend upon a sextic.   The methods of Tschirnhaus, Euler, and Bézout lead to equations of twenty-fourth degree. Lagrange sought in vain to find a resolving function which should satisfy an equation of degree less than five.

It is not uninteresting also to remark, that whenever an equation of prime degree is algebraically soluble, Lagrange's method leads us directly to the solution.   When $n = 5$, this was already noticed by Malfatti, a contemporary of Lagrange.

When n is a composite number, the foregoing considerations do not hold.

Let $n = p\nu$ where p is a prime factor.

Lagrange writes the roots in the array

$$\begin{array}{ccccccc} x_1 x_2 & . & . & . & . & . & x_\nu \\ x_{\nu+1} & . & . & . & . & x_{2\nu} \\ x_{2\nu+1} & . & . & . & . & x_{3\nu} \\ . & . & . & . & . & . & . \\ x_{(p-1)\nu+1} & . & . & . & x_{p\nu} \end{array}$$

Let the equation whose roots form the ith row be $\phi_i = 0$. The coefficients of this equation are symmetric functions of the elements of the corresponding row.

If $X_i = x_{(i-1)\nu+1} + x_{(i-1)\nu+2} + \ldots x_{i\nu}$, $i = 1, 2 \ldots p$, the coefficients of $\phi_i = 0$ are rational in X_i, and when X_1, $X_2 \ldots$ are known, the coefficients of $\phi_1 = 0$, $\phi_2 = 0 \ldots$ are rationally known, and the solution of an equation of degree $p\nu$ is reduced to the solution of p equations of degree ν. If now ν be composite we may break ν into two factors, $\nu = p_1 \nu_1$, p_1 being prime, and proceed as before.

Let us now return to the determination of the quantities X_1, $X_2 \ldots$ which are roots of an equation of prime degree, namely,

$$X = (X - X_1)(X - X_2) \ldots (X - X_p) = 0,$$

whose coefficients are roots of a rational equation of degree
$$\rho = \frac{n!}{(\nu!)^p p!}.$$

The equation $X = 0$ being prime, may be solved by the foregoing method.

When

$$n = 4, \ \rho = 3; \quad n = 6, \ \rho = 10 \text{ or } 15.$$

Here again Lagrange's methods gave him a clear insight into the reason for the success and failure of his predecessors' methods, according as $n \lesseqgtr 4$ or > 4.

Particularly instructive and important is the application Lagrange made of his methods to the equations upon which the division of the circumference of the circle into n equal parts depends,* for it would have required a far less attentive reader than Lagrange's illustrious disciple, Abel, not to have perceived what slight modifications were necessary in order to apply Lagrange's methods to the corresponding equations in the theory of elliptic functions.

The equations in question have the form

$$x^{n-1} + x^{n-2} + \ldots + x + 1 = 0$$

where we suppose n prime. Lagrange proceeds as follows: Let $n - 1 = pq$, p prime. If r be an imaginary nth root of unity and a a primitive congruence root for n, we can arrange the pq roots thus:

$$
\begin{array}{ccccc}
r & r^{a^{p}} & r^{a^{2p}} & \cdots & r^{a^{(q-1)p}} \\
r^{a} & r^{a^{p+1}} & r^{a^{2p+1}} & \cdots & r^{a^{(q-1)p+1}} \\
\cdot & \cdot \ \cdot \ \cdot & \cdot \ \cdot & \cdot \ \cdot \ \cdot & \cdot \\
r^{a^{p-1}} & r^{a^{2p-1}} & r^{a^{3p-1}} & \cdots & r^{a^{pq-1}}
\end{array}
$$

If $X_1, X_2, \ldots$ denote, as in the general case just treated, the sums of the elements of the various rows, Lagrange showed that the equation

$$X = (X - X_1) \ldots (X - X_p) = 0$$

is rational and can be algebraically solved, so that the solution of the original equation depends upon the solution of p equation of degree q, and so forth. To solve the equation $X = 0$, he employs as in the general case the resolving function

$$\theta_1 = t_1^p = (X_1 + \alpha X_2 + \ldots + \alpha^{p-1} X_p)^p.$$

But this quantity is here rationally known (α being supposed known), since, for any substitution which changes r into r^a, X_1 goes over into X_2, X_2 into X_3, etc.; thus θ_1 is unaltered. Developing and arranging according to powers of α, we have

$$\theta_1 = \xi_0 + \alpha \xi_1 + \ldots + \alpha^{p-1} \xi_{p-1},$$

* LAGRANGE: Traité de la résolution des équations numériques de tous les degrés. Paris, 1808. pp. 275–311.

where the $\mathcal{E}$'s are unchanged for r, r^a. But the $\mathcal{E}$'s being rational functions of r, r^a, r^{a^2} ..., we have, e.g.,

$$\mathcal{E}_0 = A + Br + Cr^a + \ldots + Nr^{a^{n-2}}.$$

As this is unchanged for r, r^a, we have by comparison

$$B = C = \ldots = N, \quad \text{or}$$

$$\mathcal{E}_0 = A + B(r + r^a + \ldots) = A - B;$$

that is, $\mathcal{E}_0$ is known. As θ_1 is thus known, we get

$$X_1 + \alpha X_2 + \ldots + \alpha^{p-1} X_p = \sqrt[p]{\theta_1};$$

similarly $\quad X_1 + \beta X_2 + \ldots + \beta^{p-1} X_p = \sqrt[p]{\theta_2};$

$$\cdot \quad \cdot \quad \cdot \quad \cdot \quad \cdot \quad \cdot \quad \cdot \quad \cdot$$

$$X_1 + \omega X_2 + \ldots + \omega^{p-1} X_p = \sqrt[p]{\theta_{p-1}};$$

also $\quad X_1 + \quad X_2 + \ldots + X_p = -1,$

where α, $\beta \ldots \omega$ are the $p-1$ imaginary pth roots of unity. This system of linear equations gives us X_i; for example,

$$X_1 = \frac{-1 + \sqrt[p]{\theta_1} + \sqrt[p]{\theta_2} + \ldots + \sqrt[p]{\theta_{p-1}}}{p}.$$

The roots of the equation

$$x^{n-1} + x^{n-2} + \ldots + x + 1 = 0$$

are rational functions of one of them, x_0:

(1) $\qquad x_i = \theta_i(x_0), \qquad i = 1, 2 \ldots n - 1.$

They enjoy further the property that

(2) $\qquad \theta_i \theta_\kappa x_0 = \theta_\kappa \theta_i x_0.$

But just these properties (1), (2) are enjoyed by the n^2 roots of the equation $F(x) = 0$ for dividing the argument of sn$z$, namely,

$$x_{p, q} = \text{sn}\left(\frac{z}{n} + \frac{4pK + 4q\iota K'}{n}\right),$$

when we consider the quantities k^2, sn$z$, cn$z$, dn$z$, and

$$y_{p,q} = \mathrm{sn}\left(\frac{4pK + 4q\iota K'}{n}\right),$$

as known.

How closely Abel follows Lagrange in his solution of equations of the type $F(x) = 0$ is shown by the following sketch of his method :*

Let $f(x) = 0$ be any (irreducible) equation whose roots enjoy the properties (1), (2). We may represent them by the array

$$x_0 \quad \theta x_0 \ldots \theta^{u-1}x_0 ;$$

$$x_1 \quad \theta x_1 \ldots \theta^{u-1}x_1 ;$$

$$\cdot \quad \cdot \quad \cdot \quad \cdot \quad \cdot$$

$$x_{m-1}\theta x_{m-1} \ldots \theta^{u-1}x_{m-1}.$$

Consider the equation $\phi(x) = 0$, whose roots are the elements of the first row. If we suppose the coefficients of this equation to be known, it is soluble. In fact setting with Lagrange

$$\psi_0 = (x_0 + \alpha\theta x_0 + \ldots + \alpha^{u-1}\theta^{u-1}x_0)^u,$$

ψ_0 remains unchanged for $|\,x_0,\ \theta x_0\,|$, as in the case of the cyclotomic equations just considered.

But ψ_0 is rational, for if we denote by ψ_m what ψ_0 becomes after the substitution $|\,x_0,\ \theta^m x_0\,|$, since $\psi_m = \psi_0$, we have $\psi_0 = \psi_1 = \ldots \psi_{u-1}$, whence

$$\psi_0 = \frac{1}{n}(\psi_0 + \psi_1 + \ldots + \psi_{n-1}),$$

a symmetric function of the roots of $\phi(x) = 0$.

Thus, precisely as before, we have a system of linear equations which gives, for example,

$$x_0 = \frac{-A + \sqrt[u]{\psi_0} + \sqrt[n]{\psi_1} + \ldots + \sqrt[n]{\psi_{n-1}}}{n},$$

where A is the coefficient of x^{u-1} in $\phi(x) = 0$.

We return to the equation upon which the coefficients of $\phi(x) = 0$ depend.

---

* ABEL: Mémoire sur une classe particulière d'équations résolubles algébriquements. *Crelle*, vol. 4. Also, Œuvres, 2d ed., vol. I, pp. 478–507.

Let X_1 be the oft-considered symmetric function of the elements of the first row,

$$X_1 = x_0 + \theta x_0 + \ldots + \theta^{n-1}x_0 \text{, etc.}$$

The equation

$$X = (X - X_1)(X - X_2) \ldots (X - X_m) = 0$$

is rational.   In fact

$$X_1 = F(x_0) = F(\theta x_0) = \ldots = F(\theta^{n-1}x_0).$$

Similarly $\quad X_2 = F(x_1) = F(\theta x_1) = \ldots = F(\theta^{n-1}x_1).$

$$\cdot \quad \cdot \quad \cdot \quad \cdot \quad \cdot \quad \cdot \quad \cdot$$

Hence $\quad X_1{}^\kappa = \frac{1}{n}\{(F(x_0))^\kappa + \ldots + (F(\theta^{n-1}x_0))^\kappa\};$

$$X_2{}^\kappa = \frac{1}{n}\{(Fx_1)^\kappa + \ldots + (F(\theta^{n-1}x_1))^\kappa\},$$

$$\cdot \quad \cdot \quad \cdot \quad \cdot \quad \cdot \quad \cdot \quad \cdot$$

and thus $X_1{}^\kappa + X_2{}^\kappa + \ldots + X_m{}^\kappa$ is rational.

Now the equation $X = 0$, having the same properties as the original equation $f(x) = 0$, this last is algebraically soluble, and we have the theorem that the equation upon which the division of the argument of the elliptic function $\operatorname{sn}(z)$ depends is (under the previous assumptions) algebraically soluble.

Leaving Abel now, I must pass on to a last and even more striking example of the wonderful powers of Lagrange's "*calcul*" to announce *à priori* the results which one should expect. Lagrange, as I have remarked, had vainly endeavored to find a rational resolving function for the quintic which would satisfy an equation of degree less than five, and so place one in a position to effect the solution of this celebrated equation. Ruffini, an Italian contemporary of Lagrange, and his ardent disciple, succeeded by Lagrange's own methods in proving that no such function existed; in fact he demonstrated quite generally that no rational function of n elements existed which took on three or four values for the symmetric group, n being > 4. But Ruffini was too convinced of the latent power of his great countryman's methods to stop here: he boldly undertook* by their means to prove that the alge-

---

* P. RUFFINI : Reflessioni intorno alla soluzione delle equazioni algebraiche generali.   Modena, 1813.

braical solution of the general equation of degree > 4 was impossible. Although not altogether successful in his attempt, I wish to show with what simple means he did prove that if the expression for a root can be given such a form that the radicals in it are rational functions of the roots of the given equation, then the algebraical solution is impossible when the degree of the equation surpasses four.

In fact such an expression for the root could always be arrived at as follows: Let A_1 be a rational quantity, and let n_1 be a prime; then $P_1{}^{n_1} = A_1$ defines a first irrationality. Let A_2 be any rational function of quantities originally rational and P_1; then $P_2{}^{n_2} = A_2$ defines a second irrationality. Continuing in this way, any root of the given equation has the form $x_1 = A$, where A is a rational function of $P_1, P_2 \ldots$

Let now s be the cyclic substitution $s = (1\ 2\ 3\ 4\ 5)$, and let $P_s, P_{s^2}, P_{s^3}, P_{s^4}$ be the values of P_1 for s, s^2, s^3, and s^4, respectively. Then $P_s = \beta P_1$ where $\beta^{n_1} = 1$. Operating with $s, s^2 \ldots$ this gives

$$P_{s^2} = \beta^2 P_1, \quad P_{s^3} = \beta^3 P_1, \quad P_{s^4} = \beta^4 P_1; \quad \therefore \quad \beta^5 = 1.$$

Similarly let P_σ be the value of P_1 for $\sigma = (1\ 2\ 3)$; then $P_\sigma = \gamma P_1$, and thus $\gamma^3 = 1$. But $P_{s\sigma} = \beta\gamma P_1$; hence, since $(s\sigma)^5 = 1, \beta^5\gamma^5 = 1; \therefore \gamma = 1$. Similarly, if $\rho = (3\ 4\ 5)$, then $P_\rho = P_{\rho^2} = P_1$. Now $\rho\sigma = s$; hence $P_{\rho\sigma} = P_s = P_1$. But $P_s = \beta P_1$; thus $\beta = 1$. Hence P_1 remains unaltered for s, thus also P_2, etc.; hence finally A. Thus the right hand of $x_1 = A$ is unchanged for s while the left-hand side is changed. Thus it is impossible to solve algebraically the general equation whose degree surpasses four.

The limited time at my disposal has not permitted me to discuss Lagrange's claims in detail; but the few examples I have chosen from Lagrange himself and from his immediate disciples will show, I think, how incomparably superior his methods were to those of his predecessors, Hudde, Saunderson, Le Sœur, and Waring; and it would be no difficult or ungrateful task to show how easily the ideas of Galois spring from the same source that inspired Ruffini, Cauchy, and Abel

NEW HAVEN, CONN.

ON THE RUFFINI–ABELIAN THEOREM.

BY PROFESSOR J. PIERPONT.

Gauss having rigorously established in 1799 the fundamental theorem of algebra that every equation of degree n possesses n roots real or imaginary,* it was natural to inquire more closely into the nature of these quantities. When n was less than five, it had long been known that these roots could in every case be expressed as explicit algebraic functions of the coefficients; further it was known that for every degree equations existed of more or less special nature for which this was still true. The unsuccessful attempts of the foremost mathematicians of the century which was just closing to find such expressions for the equation of degree five, when the coefficients were left indeterminate, had rendered it very doubtful if the roots of the general equation of degree greater than four possessed this property. Between the years 1799 and 1813 an Italian mathematician, Ruffini,† made several attempts to establish the justice of these doubts ; but his reasoning although highly interesting and valuable is not conclusive, and the question remained open until the publication of Abel's ‡ celebrated argument in 1826, where he proved that it was impossible to express the roots of an equation of degree greater than four, as explicit algebraic functions of the coefficients when these last were left indeterminate.

Abel's demonstration, however, is not all that could be desired. In the first place, as he was ignorant of Ruffini's beautiful researches, the substitution-theoretical part of his paper is unnecessarily roundabout; secondly, in the algebraical part Abel unnecessarily complicates his proof with a classification of functions according to order and degree. Here he commits an error which caused as acute a mind as Hamilton to declare that " it renders it difficult to judge of the validity of his subsequent reasoning." In this, however, Hamilton was misled, the classification in question being entirely superfluous here, however important in other

---

* GAUSS, Werke, vol. III. Compare also the interesting paper by Bôcher, BULLETIN, May, 1895.

† Cf. BURKHARDT. Supplement of the *Zeitschrift für Mathematik u. Physik*, vol. 37.

‡ *Oeuvres Complètes*, 2d edition, vol. I., p. 66. It is interesting to compare this with Abel's first attempt which forms the third paper of the new edition.

Reprinted from Bull. Amer. Math. Soc. 2 (April 1896), 200–221 algebraical investigations.   In any case it was highly desirable that a theorem of such far-reaching importance and which stands as it were at the very gateway of any algebraical theory of the roots of higher equations, should be demonstrated in a simple, direct and rigorous manner. Such a demonstration was published by Kronecker in the Monatsberichte* of the Berlin Academy for 1879.   This paper has not received the attention of writers of treatises on algebra which it deserves; such standard modern treatises† as Serret, Petersen, Carnoy, all contain demonstrations of the theorem which are imperfect.   Serret's demonstration is in its algebraical part almost a word for word reproduction of Abel's memoir and thus contains the unfortunate error noted by Hamilton; the demonstration given by Petersen and Carnoy is vitiated by the same tacit postulate made by Ruffini in the Modena treatise of 1813.

Since the publication of Kronecker's paper in 1879, no other proof has appeared, and yet it seems to me that Kronecker in following Abel too closely in the substitution-theoretical part of his demonstration has not given the simplest proof possible; on the other hand, in the algebraical part, in condensing to the utmost the demonstration, he presupposes that his readers are familiar with the form of proof given by Abel.   I propose then in the following lines to give a demonstration of the Ruffini-Abelian Theorem which shall be as direct and *self-contained* as possible ; in doing this I hope to render a service to a large class of readers whose studies have led them away from algebraical theories, but who would be glad to see this celebrated theorem demonstrated without implying more or less familiarity with the higher parts of algebra.   To aid in this I give at the end of this paper § 8 an illustrative example worked out in detail.   In addition to this demonstration I give two others: one, a modification of Ruffini's form; the second, Kronecker's modification of Abel's form.   As will be observed, the three forms of demonstration have their algebraical part, essentially due to Abel, in common, while the substitution-theoretical part increases in complexity.

§ 1

We begin by giving a few definitions and establishing one or two elementary theorems.

---

* p. 205 *Vereinfachung des Abel'schen Beweises*, etc.

† SERRET : *Algèbre supérieure* 5th ed.   PETERSEN : *Theorie der algebraischen Gleichungen.*   CARNOY : *Cours d'Algèbre supérieure.*

We say $f(\lambda, \mu, \nu \cdots)$ is an integral function of the quantities $\lambda, \mu, \nu, \cdots$ when its calculation requires only the operations of addition, subtraction and multiplication in regard to these quantities; every such function is of the type $\Sigma C \lambda^a \mu^b \nu^c \cdots$ where the coefficients C do not involve $\lambda, \mu, \nu \cdots$ and the exponents $a, b, c \cdots$ are positive integers. The function $f(\lambda, \mu, \nu \cdots)$ is said to be a rational function of $\lambda, \mu, \nu \cdots$ when its calculation requires only the operations of addition, subtraction, multiplication and division in regard to them; every such function is the quotient of two integral functions of $\lambda, \mu, \nu \cdots$ and is thus of the type

$$f(\lambda, \mu, \nu \cdots) = \frac{f_1(\lambda, \mu, \nu \cdots)}{f_2(\lambda, \mu, \nu \cdots)}$$

f_1, f_2, being integral. A function $f(a, b, c, \cdots; \lambda, \mu, \nu \cdots)$ may be an integral function in regard to certain quantities $\lambda, \mu, \nu \cdots$ and rational in regard to certain others, $a, b, c \cdots$; if f involves only the quantities $a, b, c \cdots \lambda, \mu, \nu \cdots$ and certain rational numbers, we say $f(a, b, c \cdots \lambda, \mu, \nu \cdots)$ is an integral function of $\lambda, \mu, \nu \cdots$ whose coefficients are rational functions of $a, b, c \cdots$ with rational-number coefficients, or more shortly f is an integral function of $\lambda, \mu, \nu \cdots$ with coefficients rational in $a, b, c \cdots$.

Let $\varphi(\xi, a, b, c \cdots)$ be an integral function of ξ whose coefficients are rational in $a, b, c \cdots$; if $\xi_1, \xi_2, \xi_3 \cdots$ be the values of ξ for which $\varphi = 0$,

$$\varphi(\xi) = A(\xi - \xi_1)(\xi - \xi_2)(\xi - \xi_3) \cdots;$$

similarly if $\eta_1 \eta_2 \eta_3 \cdots$ be the zeros of $\psi(\eta, a', b', c' \cdots)$

$$\psi(\eta) = B(\eta - \eta_1)(\eta - \eta_2)(\eta - \eta_3) \cdots$$

If some of the η's are equal to some of the ξ's as

$$\eta_1 = \xi_1, \ \eta_2 = \xi_2, \cdots \eta_\mu = \xi_\mu$$

then φ and ψ have a common divisor

$$\delta = (\xi - \xi_1) \cdots (\xi - \xi_\mu)$$

As δ can be obtained by the process of finding the greatest common divisor, and as this process never requires other than rational operations on the coefficients of φ, ψ which are rational in the quantities $a, b, c \cdots a', b', c'$, it follows that δ is an integral function of ξ with coefficients rational in $a, b, c \cdots a', b', c' \cdots$. The function $f(\xi, a, b, c \cdots)$ is said to be irreducible in regard to certain quantities $\lambda, \mu, \nu \cdots$ when it is not the product of factors whose coefficients are rational in $\lambda, \mu, \nu, \cdots$; in the same way the equation $f(\xi) = 0$ is said to be irreducible under the same circumstances. From this definition it follows that if $f(\xi, a, b, c \cdots)$ be irreducible in regard to $a, b, c \cdots$ and certain quantities $\lambda, \mu, \nu \cdots$ and have a divisor δ in common with $\varphi(\xi, a, b, c \cdots \lambda, \mu, \nu \cdots)$ then φ is divisible by f. For δ as we saw is of the form $\delta(\xi, a, b, c \cdots \lambda, \mu, \nu \cdots)$; which being a factor common to f and φ, requires that f possess a factor whose coefficients are rational in $a, b, c \cdots \lambda, \mu, \nu \cdots$ which is impossible unless $f = \delta$, that is unless φ is divisible by f. Hence:

THEOREM I : If

$$f(\xi, a, b, c \cdots) = \xi^m + A_1 \xi^{m-1} + \cdots + A_m = 0$$

be irreducible in regard to $a, b, c \cdots \lambda, \mu, \nu \cdots$, and if at the same time ξ satisfy the equation of *less* degree

$$\xi^\mu + B_1 \xi^{\mu-1} + \cdots + B_\mu = 0 \qquad\qquad \mu < m$$

where the B are rational in $a, b, c \cdots \lambda, \mu, \nu \cdots$ then

$$B_1 = 0 \quad B_2 = 0 \cdots \cdots B_\mu = 0$$

We turn now to the substitution theory. The number of permutations of n things

$$1, 2, 3 \cdots \cdots n$$

is $n!$; let

$$l_1 \ l_2 \ l_3 \cdots \cdots l_n$$

denote one of these permutations; the operation which replaces 1 by l_1, 2 by $l_2 \cdots n$ by l_n is called a substitution and is denoted by the symbol

$$s = \begin{pmatrix} 1 & 2 & 3 \cdots n \\ l_1 & l_2 & l_3 \cdots l_n \end{pmatrix}$$

where the order of the top row is indifferent if only the l_1 is under 1, l_2 under 2 etc., in each case. As to each of the $n!$ permutations corresponds a substitution, there are $n!$ substitutions affecting the n things $1, 2 \cdots n$. I shall speak of them as the substitutions of the symmetric group G or as the substitutions of G. One of these substitutions is

$$\begin{pmatrix} 1 & 2 & 3 \cdots n \\ 1 & 2 & 3 \cdots n \end{pmatrix}$$

which leaving each element unchanged is called the identical substitution and denoted briefly by 1. Let $s, t, u, \cdots$

be certain substitutions, the operation arising from operating first with s, then with t is denoted by $s.t$ and is called
the product of s, t; similarly $s\,t\,u$ is the resulting operation
arising from operating first with s, then with t finally with u.

A substitution which replaces each element of the top row
by the following, finally the last by the first is called a cycle
or a circular substitution ; thus

$$\begin{pmatrix}1\;2\;3\;4\\2\;3\;4\;1\end{pmatrix}, \quad \begin{pmatrix}4\;2\;1\;5\;3\\2\;1\;5\;3\;4\end{pmatrix}$$

are such.   Every substitution is the product of cycles, thus

$$\begin{pmatrix}1\;2\;3\;4\;5\;6\;7\\3\;1\;7\;5\;6\;4\;2\end{pmatrix}=\begin{pmatrix}1\;3\;7\;2\\3\;7\;2\;1\end{pmatrix}\begin{pmatrix}4\;5\;6\\5\;6\;4\end{pmatrix}$$

$$\begin{pmatrix}1\,2\,3\,4\,5\,6\,7\,8\,9\,10\,11\\10\,8\,3\,6\,7\,5\,4\,9\,2\,1\;\;11\end{pmatrix}=\begin{pmatrix}1\,10\\10\,1\end{pmatrix}\begin{pmatrix}2\,8\,9\\8\,9\,2\end{pmatrix}\begin{pmatrix}3\\3\end{pmatrix}\begin{pmatrix}4\,6\,5\,7\\6\,5\,7\,4\end{pmatrix}\begin{pmatrix}11\\11\end{pmatrix}$$

A circular substitution of two elements as (12) is a transposition.   To every substitution s corresponds *one* substitution t such that $st=1$; thus if

$$s=\begin{pmatrix}1\;2\;3\;4\\4\;3\;1\;2\end{pmatrix}$$

the substitution t is

$$t=\begin{pmatrix}4\;3\;1\;2\\1\;2\;3\;4\end{pmatrix}$$

for s converts

$$1\;2\;3\;4$$
into
$$4\;3\;1\;2$$

respectively but each element of this last row is converted by
t into

$$1\;2\;3\;4$$

respectively, so the joint effect of s and t, or $s.t$ is to
leave each element unchanged $\therefore st=1$.   The substitution t
is called the inverse of s and denoted by s^{-1}, so that $s\,s^{-1}=1$.
For brevity also we put $s.s=s^2$, $s.s.s=s^3$, etc.

Let $x_1\,x_2\cdots x_n$ be the roots of $x^n+a_1\,x^{n-1}+\cdots+a_n=0$, and
let $\varphi\,(x_1, x_2\cdots x_n;\; \lambda,\, \mu,\cdots)$ be a rational function of $x_1\,x_2\cdots x_n$
and certain quantities $\lambda,\, \mu,\cdots$; if φ remain invariant *in form*,
however we permute the indices 1, 2, 3$\cdots n$, or what is the
same thing, for all the substitutions of the symmetric group
G, φ is said to be a rational symmetric function of $x_1\,x_2\cdots x_n$.

An elementary theorem of algebra is the following :

THEOREM II.   Every rational symmetric function $\varphi\,(x_1\,x_2, \cdots x_n\,;\,\lambda, \mu, \nu)$ of the roots $x_1\,x_2 \cdots x_n$ of the equation $x^n + a_1\,x^{n-1} + \cdots + a_n = 0$, is a rational function of the coefficients $a_1, a_2 \cdots a_n$ and the quantities $\lambda, \mu, \nu \cdots$.

The function $f\,(\lambda, \mu, \nu \cdots .)$ is said to be an explicit algebraical function of $\lambda, \mu, \nu \cdots$ when its calculation requires in addition to the rational operations only the extraction of roots in regard to the quantities $\lambda, \mu, \nu \cdots$.

We shall call the equation

$$(a) \qquad f\,(x) = x^n + a_1\,x^{n-1} + \cdots + a_n = 0$$

general when its coefficients are independent variables, and it shall possess an algebraic solution when their exists an explicit algebraic function of the coefficients which satisfies it. We propose to prove that when n is greater than four, no such function exists by showing that the admission of its existence leads to a contradiction.

§ 2

In fact let us seek the most general form of an explicit algebraical function of the coefficients $a_1\,a_2 \cdots a_n$. It may be obtained as Ruffini remarked as follows :

Taking a rational function of the a's as $f_1(a_1\,a_2 \cdots a_n)$ we extract a p_1^{th} root ; calling this radical R_1 we form a rational function of the a's and of R_1 as $f_2(a_1 \cdots a_n, R_1)$ and extract a p_2^{th} root ; calling this radical R_2 we form a rational function of the a's and of R_1 and R_2 as $f_3(a_1\,a_2 \cdots a_n\,;\,R_1\,R_2)$, and extract a p_3^{th} root which we call R_3 : proceeding in this way every explicit algebraical function x_n will be defined by the *suite* of equations :

$$(A) \quad \begin{aligned} & R_1{}^{p_1} = f_1\,(a_1\,a_2 \cdots a_n),\ R_2{}^{p_2} = f_2\,(a_1 \cdots a_n\,;\,R_1), \cdots \\ & R_\lambda{}^{p_\lambda} = f_\lambda\,(a_1 a_2 \cdots a_n\,;\,R_1\,R_2 \cdots R_{\lambda-1}),\ x_0 = f(a_1 a_2 \cdots a_n, R_1, R_2 \cdots R_\lambda) \end{aligned}$$

the functions f being all rational in regard to the quantities within the parentheses.

We proceed now to show how the *suite* may be simplified without losing its generality. In the first place the exponents p_k may be taken to be primes, since the extraction of an mn^{th} root may be effected by the extraction of an mth, and from that an nth root. Another simplification arises by observing that no loss of generality is incurred by considering f_k to be an integral function of the radicals $R_1\,R_2 \cdots R_{-1k}$ with coefficients rational in $a_1, a_2, \cdots a_n$.

For let for example $f(a_1 \cdot a_n, u)$ be a rational function of u with coefficients rational in $a_1 \cdots a_n$, and

(1) $$u^p = g\,(a_1 \cdots a_n)$$

then f is of the form

(2) $$f = \frac{b_0 + b_1 u + b_2 u^2 + \cdots + b_{p-1} u^{p-1}}{c_0 + c_1 u + c_2 u^2 + \cdots + c_{r-1} u^{p-1}} = \frac{\varphi\,(u)}{\psi\,(u)}$$

the coefficients b, c being rational in the a's. It is to be noticed that neither φ nor ψ contains powers of u higher than $p-1$, since all higher powers may be reduced by observing that

$$u^p = g \quad u^{p+1} = u^p.u = g.u, \quad u^{p+2} = u^p.u^2 = g\,u^2 \cdots$$

and g is rational in the a's.

The expression (2) may be made integral in regard to u as follows: let the p roots of (1) be

$$u, \quad u_1 = \omega\,u \quad u_2 = \omega^2 u \cdots u_{p-1} = \omega^{p-1} u$$

where $\omega\ \omega^2 \cdots$ are the roots of the equation

(3) $$\frac{x^p - 1}{x - 1} = x^{r-1} + x^{r-2} + \cdots + x + 1 = 0$$

The product

$$P = \psi\,(u)\,\psi\,(u_1) \cdots \psi(u_{p-1})$$

is a symmetric function of the roots of (1) and is thus a rational function $P(a_1 \cdots a_n)$ of $a_1 \cdots a_n$; further the product

$$Q = \psi(u_1)\psi(u_2) \cdots \psi(u_{p-1})$$

considered as a function of ω is symmetric in the roots of (3) and is thus an integral function $Q\,(a_1, \cdots a_n;\, u)$ of u with coefficients rational in $a_1 \cdots a_n$. We have thus, multiplying numerator and denominator by Q,

$$f = \frac{\varphi\,Q}{\psi\,Q} = \frac{\varphi\,Q}{P};$$

and the right hand member is now an integral function of u with coefficients rational in $a_1 \cdots a_n$. Thus every rational function $f\,(a_1 \cdots a_n, u)$ can be expressed as an integral function of u, with coefficients rational in $a_1\,a_2 \cdots a_n$.

Precisely the same reasoning applies to a rational function $f\,(a_1\,a_2 \cdots a_n, u, v)$ where

(4) $$v^q = h\,(a_1 \cdots a_n, u)$$

the function h being an integral function of u with coefficients rational in $a_1 a_2 \cdots a_n$. In fact arranging according to ascending power of v

$$f(u, v) = \frac{l_0 + l_1 v + l_2 v^2 \cdots + l_{q-1} v^{q-1}}{m_0 + m_1 v + m_2 v^2 + \cdots + m_{q-1} v^{q-1}} = \frac{\lambda\,(v)}{\mu\,(v)}$$

where the coefficients l, m, as rational functions of $a_1 \cdots a_n$, u may be taken as integral functions of u with coefficients rational in $a_1 \cdots a_n$. Making the denominator $\mu(v)$ symmetric in respect to the roots

$$v,\ v_1,\ v_2 \cdots\cdots\ v_{q-1}$$

of (4) by multiplying numerator and denominator by $\mu\,(v_1) \cdots \mu\,(v_{q-1})$ we see that

$$f(u, v) = \frac{\lambda\,(v)\ \mu\,(v_1)\ \cdots\ \mu\,(v_{q-1})}{\mu\,(v)\ \mu\,(v_1)\ \cdots\ \mu\,(v_{q-1})}$$

is an integral function of u, v, with coefficients rational in $a_1 \cdots a_n$. These considerations being applicable to rational functions of any number of radicals, we see that without loss of generality we may assume that the functions f in (A) are integral functions in regard to $R_1 R_2 \cdots$ with coefficients rational in respect to $a_1 a_2 \cdots a_n$.

Finally we observe that we can assume that the equations

(5) $$R_k^{p_k} = f_k \qquad k = 1, 2 \cdots \lambda$$

are irreducible in respect to

(6) $$a_1 \cdots a_n;\ R_1,\ R_2 \cdots R_{k-1}$$

For if (5) be reducible let

(7) $$(X - R_k)(X - R_k') \cdots (X - R_k^{(\mu-1)})$$

be a factor rational in respect to the quantities (6), R_k, $R_k' \cdots R_k^{(\mu-1)}$ being roots of (5); then the absolute term of (7) or

$$R_k\, R_k' \cdots R_k'^{(\mu-1)} = \omega\, R_k^\mu, \qquad \omega^{p_k} = 1$$

is rational in (6), or if λ satisfy $\lambda \mu \equiv 1 \bmod p_k$, $\omega^\lambda R_k$ is rational in (6); that is R_k is rational in ω and the quantities (6). But as the determination of ω, as is well known, depends upon the solution of a *suite* of binomial equations whose degrees are the prime factors of $p_k - 1$, the assumption that the equations of the *suite* (A) are irreducible in the above sense, involves no loss of generality.

Thus to sum up: every explicit algebraical function of $a_1 a_2 \cdots a_n$ is represented by the *suite* (A) where f_k is an integral function of the radicals $R_1, R_2 \cdots R_{k-1}$ whose coefficients are rational in $a_1 a_2 \cdots a_n$; the equations $R_k^{p_k} = f_k$ are irreducible in respect to $a_1 \cdots a_n$; $R_1 \cdots R_{k-1}$; the exponents p_k are primes, finally

(b) $x_0 = \Sigma A_h R_\lambda^h$ $h = 0, 1 \cdots p_\lambda - 1$

where A_h is an integral function of $R_1 \cdots R_{\lambda-1}$, with coefficients rational in $a_1 \cdots a_n$.

§ 3

We impose now the condition that the explicit algebraical function x_0 shall satisfy the equation (a); this will give us a relation between the roots $x_0 x_1 \cdots$ and the radical R_λ which will indicate a method of simplifying the suite (A) still further. Raising x_0 to the various powers, and substituting these values of $x_0, x_0^2, x_0^3, \cdots x_0^n$ in (a), we have the equation.

(1) $0 = B_0 + B_1 R_\lambda + B_2 R_\lambda^2 + \cdots + B_{p_\lambda - 1} \; R_\lambda^{p_\lambda - 1}$

and making use of

(2) $R^{p_\lambda} = f_\lambda$

to reduce powers of R_λ higher than $p_{\lambda-1}$; the coefficients B are rational in

$$a_0 a_1 \cdots a_n, R_1, R_2 \cdots, R_{\lambda-1}$$

and hence by theorem I., the coefficients B all vanish.* Thus (1) is satisfied when we replace R_λ by $a R_\lambda$ where $a^p_\lambda = 1$. Hence the expressions obtained from (b) when we replace there R_λ by $a^k R_\lambda$, that is

(3) $x_k = \Sigma A_h a^{hk} R_\lambda^h$ $k = 0, 1 \cdots p_\lambda - 1$

satisfy (1) and are thus roots of (a). Here all the coefficients A cannot vanish, let A_h be one of these; if we multiply both sides of the p_λ equations (3) respectively by a^{-h}, $a^{-2h} \cdots a^{-(p_\lambda - 1)h}$, and add, we get, remembering that

$$1 + a^r + a^{2r} + \cdots + a^{\overset{(p-1)r}{\lambda}} = 0$$

the equation

(4) $\Sigma a^{-hk} x_k = p_\lambda A_h R_\lambda^h$

which gives us the relation in question.

---

* We have assumed the *suite* to embrace only irreducible equations; in practice it is convenient to defer the elimination of reducible elements till this stage where we have the means of deciding concerning reducibility.

This relation shows us how we may express x_o as an integral function of certain radicals

$$P_1, P_2 \cdots\cdots P_\lambda$$

where these radicals enjoy the peculiar property of being integral functions of $x_o \, x_1 \cdots\cdots x_{n-1}$ and certain roots of unity $a, \beta \cdots \omega$ with rational-number coefficients.   In fact put†

$$(5) \qquad\qquad P_\lambda = p_\lambda A_\lambda R_\lambda^h;$$

this equation together with (2) shows that R_λ is a rational function of P_λ and the quantities $R_1, \cdots R_{\lambda-1}$, since the assumption that (2) is irreducible in respect to $R_1 \cdots R_{\lambda-1} P_\lambda$ is, on account of (3) inadmissible.   (Theorem I). As R_λ is rational in respect to these quantities so is x_o. That P_λ satisfies a binomial equation of the type

$$P_{\lambda\lambda}^p = g_\lambda (a_1 \cdots a_n, R_1 R_2 \cdots R_{\lambda-1})$$

is evident ; for

$$P_{\lambda\lambda}^p = p_{\lambda\lambda}^p A_{\lambda\lambda}^p R_{\lambda\lambda}^{p\,h} = p_{\lambda\lambda}^p A_{\lambda\lambda}^p f_\lambda^h$$

which is a rational function of $a_1 \cdots a_n \, R_1 \cdots R_{\lambda-1}$ which we denote by g_λ.

The radical R_λ being replaced by P_λ, we proceed now to show how we may replace $R_{\lambda-1}$ by $P_{\lambda-1}$.   Put

$$\xi_0 = P_{\lambda\lambda}^p = (\Sigma a^{-\lambda\,k} x_k)^p{}_\lambda$$

and let $\xi_0, \xi_1, \xi_2 \cdots$ be the n ! functions (different or not) arising from applying the n ! substitutions of the symmetric group to ξ_0.   Form now the equation

$$\Phi(\xi) = (\xi - \xi_o)\,(\xi - \xi_1) \cdots\cdots = 0,$$

its coefficients being symmetric in $x_o, x_1, \cdots x_n$ are rational in the coefficients $a_1 \cdots a_n$.

Now a moment's reflection shows that the reasoning applied to the equations

$$f(x) = 0, \qquad x_0 = f_\lambda (a_1 \cdots a_n; \; R_1 \cdots R_\lambda)$$

---

† This step could obviously be omitted if for $h = 1$, A_λ is a rational number.

can be applied word for word to the equations

$$\Phi(\xi)=0, \qquad \xi_0=g_\lambda\,(a_1\cdots a_n;\; R_1\cdots R_{\lambda-1})$$

In fact referring to the reasoning just made in regard to x_0, R_λ we see that we may write

(b')
$$\xi_0=\Sigma\,A_r'\,R_{\lambda-1}^r \qquad r=0,1\cdots p_{\lambda-1}1.$$

which is the equation analogous to (b).

This value of ξ_0 substituted in $\Phi=0$ gives

(1')
$$0=B_0'+B_1'R_{\lambda-1}+\cdots B'_{p-1}\,R_{\lambda-1}^{\,\prime-1}$$

which is analagous to (1). This shows that the quantities

(3') $\quad \xi_s=\Sigma\,A_r'\,\beta^{rs}\,R_{\lambda-1}^r \qquad s=0,1\cdots p_{\lambda-1}-1, \qquad \beta^{p_{\lambda-1}}=1$

also satisfy $\Phi=0$. These equations now give as before the relation

(4') $\qquad\qquad \Sigma\,\beta^{-rs}\,x_s=p_{\lambda-1}\,A_r'\,R_{\lambda-1}^r$

analogous to (4). It follows thus, if we put

$$P_{\lambda-1}=p_{\lambda-1}\,A_r'\,R_{\lambda-1}^r$$

that $P_{\lambda-1}$ is a rational function of $R_1, R_2, \cdots R_{\lambda-2}$ and $P_{\lambda-1}$ while on the one hand $P_{\lambda-1}$ satisfies a binomial equation of the type

$$P_{\lambda-1}^{p_{\lambda-1}} = g_{\lambda-1}\,(a_1\cdots a_n;\; R_1, R_2, R_{\lambda-2})$$

and on the other is an integral function of $\xi_0\,\xi_1\cdots$ and β. But $\xi_0\,\xi_1\cdots$ are integral functions of $x_0\,x_1\cdots$ and a, hence $P_{\lambda-1}$ is an integral function of $x_0\,x_1\cdots x_{n-1}$ and the roots of unity a, β; the coefficients of these functions being rational numbers. Returning now to the expression for x_0, namely :

$$x_0 = f\,(a_1\cdots a_n,\, R_1\cdots R_{\lambda-1},\, R_\lambda),$$

we see that we are in the position to replace the radicals $R_{\lambda-1}, R_\lambda$ by the radicals $P_{\lambda-1}, P_\lambda$. The same reasoning being applicable to the remaining radicals we have the following :

THEOREM III. If an equation is algebraically solvable, we can always give the explicit algebraic expression for a root such a form that all the radicals entering it are integral functions of the roots $x_0\,x_1\cdots x_{n-1}$ and certain roots of unity $a, \beta,\cdots\omega$, the coefficients of these functions being rational numbers;

the radicals themselves are determined by a *suite* of be
nomial equations

$$(B) \qquad P_1^{p_1} = g_1(a_1 \cdots a_n) P_2^{p_2} = g_2(a_1 \cdots a_n; P_1) \cdots\cdots P_\lambda^{p_\lambda} = $$
$$g_\lambda (a_1 \cdots a_n; P_1 P_2 \cdots P_{\lambda-1})$$

while x_0 is given by

$$(C) \qquad\qquad x_0 = g (a_1 \cdots a_n; P_1 \ P_2 \cdots P_\lambda),$$

the functions g being integral functions of $P_1 \cdots\cdots P_\lambda$ with
coefficients rational in $a_1 \cdots a_n$.

§4.

The theorem just proved affords us a sure foundation to
construct the substitution-theoretical part of the demonstration; we give first a form due to Ruffini.* Consider the
rational integral function of the roots†

$$\varphi (x_1 \ x_2 \cdots x_n)$$

defined by the first equation of (B); let $\varphi_s, \varphi_{s^2} \cdots\cdots$ be the
values of φ after application of $s \ s^2 \cdots$ where

$$s = \begin{pmatrix} 1 \ 2 \ 3 \ 4 \ 5 \\ 2 \ 3 \ 4 \ 5 \ 1 \end{pmatrix},$$

then since $g_1 (a_1 \cdots a_n)$ is a symmetric function of the roots
it remains unchanged for s. Applying then s to the identity

$$\varphi^{p_1} = g_1 (a_1 \cdots a_n)$$

it becomes $\quad \varphi_s^{p_1} = g_1 \quad$ whence

$$(1) \qquad\qquad \varphi_s = \beta \, \varphi \quad \text{where } \beta^{p_1} = 1$$

Applying s to the identity (1) it becomes

$$\varphi_{s^2} = \beta \varphi_s = \beta^2 \varphi$$

similarly $\varphi_{s^3} = \beta^3 \varphi$, $\varphi_{s^4} = \beta^4 \varphi$ and $\varphi_{s^5} = \beta^5 \varphi$; but as $s^5 = 1$, $\varphi_{s^5} = \varphi$
and $\beta^5 = 1$.
In the same way, if

$$\sigma = \begin{pmatrix} 1 \ 2 \ 3 \\ 2 \ 3 \ 1 \end{pmatrix}$$

---

*Cf. my paper in *Monatshefte für Mathematik und Physik*, vol. 6,
for fuller information concerning this form of proof used by Ruffini in
the Modena work already referred to.
† We designate these now by the indices 1, 2...n, as more convenient.

$\varphi_\sigma = \gamma \varphi$ where, as $\sigma^3 = 1$, $\gamma^3 = 1$

Hence the application of σ to the identity (1) gives

$$\varphi_{s\sigma} = \beta \varphi_\sigma = \beta \gamma \varphi$$

whence we conclude as above that $(\beta\gamma)^5 = 1$ since $(s\sigma)^5 = 1$. Similarly if

$$\rho = \begin{pmatrix} 3 & 4 & 5 \\ 4 & 5 & 3 \end{pmatrix}$$

then $\varphi_\rho = \varphi_{\rho^2} = \varphi$; but since $\rho\sigma = s$, $\varphi_{\rho\sigma} = \varphi_s = \varphi$; hence $\beta = 1$ since $\varphi_s = \beta\varphi$. That is φ remains not only unaltered for σ and ρ, but also for s. Consider now the rational function of the roots

$$\zeta' (x_1 x_2 \cdots x_n)$$

defined by the second equation of the *suite B*. The right hand side of the identity

$$\zeta'^{p_2} = g_2 (a_1 \cdots a_n; \varphi)$$

remaining unaltered for the substitutions s, σ, ρ we can reason in regard to ζ' precisely as we did concerning φ: hence ζ' is unchanged for these substitutions. Proceeding in this way we see that all the radicals $P_1 P_2 \cdots P_\lambda$ remain unaltered for s; but applying s to the identical relation (C) viz:

$$x_1 = g (a_1 \cdots a_n; P_1 P_2 \cdots P_\lambda)$$

we observe that the right hand side remains unaltered whereas the left hand side does not which is an absurdity. The assumption that there exist an explicit algebraical function of the coefficients $a_1 a_2 \cdots a_n$ which satisfies (A), leads thus to a contradiction.

§5

An interresting modification of Ruffini's proof is the following: We just saw that P_ν $(\nu = 1, 2)$ remain unaltered by the circular substitution $\begin{pmatrix} 1 & 2 & 3 \\ 2 & 3 & 1 \end{pmatrix}$; as the indices $1, 2, 3, 4, 5$ employed in showing this were any five of the n indices, it follows that P_ν remain unaltered for every circular substitution of three indices. These substitutions we shall call the substitutions of the alternate group which we denote shortly by Γ. We propose to show that P_ν remain unaltered for no other substitutions of the symmetric group G; having established this it follows that the radical P_2 is rational in P_1 which violates the hypothesis made in regard to the *suite* B. In the first place P_1 must change its value for every transposition $\tau = (\lambda, \mu)$ for if it remain unaltered for *any* transposition we show that it remains unaltered for *every* transposition; as every substitution is the product of

cycles and as $\begin{pmatrix} 1\ 2\ 3 \cdots m \\ 2\ 3\ 4 \cdots 1 \end{pmatrix}$ is the product of transpositions $(1\ 2)$

$(1\ 3) \cdots (1\ m)$, P_1 would remain unaltered for every substitution, and then being symmetric functions of the roots, it would be rational in $a_1\ a_2 \cdots a_n$.

Suppose then P_1 remain unaltered for τ, let λ be any substitution of G, then P_1 remains unaltered for $\varphi = \lambda^{-1}\tau\lambda$; for since $P_{\lambda^{-1}\tau} = P_{\lambda^{-1}}$ we have $P_{\lambda^{-1}\tau\lambda} = P_{\lambda^{-1}\lambda} = P_1$

Now, by properly choosing λ, ρ can represent any transposition $(l_\lambda\ l_\mu)$; for let

$$\lambda = \begin{pmatrix} 1\ 2\ 3 \cdots n \\ l_1\ l_2\ l_3 \cdots l_n \end{pmatrix}$$

then λ^{-1} converts

(1) $\qquad\qquad l_1\ l_2\ l_3 \cdots l_\lambda \cdots l_\mu \cdots l_n$

respectively into

$$1\ 2\ 3 \cdots \lambda, \cdots \mu \cdots n$$

which τ converts respectively into

$$1\ 2\ 3 \cdots \mu \cdots \lambda \cdots n$$

which finally λ converts into

(2) $\qquad\qquad l_1\ l_2\ l_3 \cdots l_\mu \cdots l_\lambda \cdots l_n\ ;$

that is the substitution $\rho = \lambda^{-1}\tau\lambda$ converts (1) into (2) so that ρ leaving every index except l_λ, l_μ unaltered and merely transposing these

$$\rho = (l_\lambda, l_\mu)$$

Thus every transposition alters P_1; but every pair of transpositions leaves it unchanged, since every such pair is of the type

$$(1\ 2)\ (3\ 4)\ \text{ or }\ (1\ 2)\ (2\ 3),$$

but the first being equal to $\begin{pmatrix} 1\ 2\ 3 \\ 2\ 3\ 1 \end{pmatrix}\begin{pmatrix} 3\ 1\ 4 \\ 1\ 4\ 3 \end{pmatrix}$ and the second

equal to $\begin{pmatrix} 1\ 3\ 2 \\ 3\ 2\ 1 \end{pmatrix}$, both are substitutions of Γ. From this we conclude that if any substitution s be the product of m transpositions, it will alter or leave unaltered the radical P_1 according as m is odd or even. This shows that however s be decomposed into a product of transpositions, their number is always odd or always even, and we say s is an odd or even substitution according as m is. As $\varGamma$ contains all the even substitutions of G, P_1 and hence P_2 also is altered for every substitution not in $\varGamma$. The number of substitutions $\varGamma$ in G is $\frac{1}{2} n!$; for let

(1) $$ s_1 = 1, \, s_2, \, s_3 \cdots s_m $$

be the even substitution, and

$$ t_1 \, t_2 \, t_3 \cdots t_\mu $$

be the odd; then,

(2) $$ t_1, \, s_2 t_1, \cdots \cdots s_m \, t_1 $$

are distinct and odd, hence $\mu \geq m$; also

$$ t_1 \cdot t_1, \; t_2 \, t_1, \; t_3 \, t_1, \cdots t_\mu \, t_1 $$

are distinct and even, hence $m \geq \mu$ $\therefore$ $m = \mu = \frac{1}{2} n!$ This shows that P_ν acquire but two values for the substitutions of G; for they remain unchanged for (1) while they both change for the substitutions (2) into P_ν', and there are no other substitutions in G.

We now conclude easily that P_2 is rational in P_1. As P_1 is root of a binominal equation, it follows that $P_1' = -P_1$; also the coefficients of

$$ \varPhi(x) = (P_2 + P_2') \, x + P_1 \, (P_2 - P_2') $$

being unaltered for the substitutions of G, are rational in $a_1 \, a_2 \cdots a_n$; denoting them by A, B, we have

$$ \varPhi(x) = A x + B $$

But $$ \varPhi(P_1) = A P_1 + B = 2 P_1 P_2, $$

hence $$ P_2 = \frac{A P_1 + B}{2 P_1} $$

and P_2 is rational in P_1 and the coefficients $a_1 \cdots a_n$.

§ 6.*

We turn now to Kronecker's form of Abel's proof. From the preceding paragraph we take as proven that P_1 is un-

---

* In anticipation of questions of priority, I remark that this paragraph formed part of a paper I had the honor to read before the Yale Mathematical Club, December 4, 1894. No changes except of the most insignificant nature have been made in the text.

altered for the substitution of Γ, and that every rational function of the roots unaltered for Γ is rational in P_1. Consider the rational function of the roots $\varphi\,(\,x_1\,x_2\cdots x_n\,)=P_2$ defined by the second equation of the *suite* (B)

$$(1) \qquad \varphi^{\,p_2} = g_2\,(P_1).$$

The $\tfrac{1}{2}n!$ functions (different or not)

$$(2) \qquad \varphi\;\varphi_1\;\varphi_2\;\cdots\cdots$$

which arise from φ on applying to it the $\tfrac{1}{2}n!$ substitution of Γ must all be of the form

$$\omega\varphi, \qquad \omega^{p_2} = 1$$

since the right hand side of (1) remains unaltered for these substitutions. Thus if φ remain unaltered for one substitution

$$\sigma = \left(\begin{matrix} 1,\,2,\,3\cdots n \\ l_1\,l_2\,l_3\cdots l_n \end{matrix}\right)$$

of Γ besides the identical substitution, all the conjugate functions (2) must also remain unaltered for σ. But such a substitution must exist since $\tfrac{1}{2}n!>p_2$; in fact the coefficients of the equation

$$(3) \qquad \Phi\,(x)=(x-\varphi)\,(x-\varphi_1)\,(x-\varphi_2)\cdots$$

being symmetric in (2) are unchanged for the substitutions of Γ and are thus rational in P_1.

But $\Phi=0$ having the root $x=\varphi$ in common with the irreducible equation

$$(4) \qquad P_2^{p_2} = g_2\,(P_1)$$

is satisfied by all the roots of (4) and hence $\tfrac{1}{2}n! \gtreqless p_2$; as p_2 is prime $\tfrac{1}{2}n!>p_2$.

We now show that the conjugate functions (2) remaining unaltered by σ a substitution different from the identical substitution, remain unaltered for every substitution

$$s = \left(\begin{matrix} 1\;2\cdots n \\ h_1\,h_2\cdots h_n \end{matrix}\right) \text{ of } \Gamma \text{ and } P_2=\varphi \text{ is thus rational in } P_1 \text{ which}$$

is contrary to hypothesis. To do this Kronecker remarks that φ remains unaltered for every substitution ρ of Γ which changes

$$l_k \text{ into } h_k \qquad k=1,2\cdots n$$

In fact $\rho = \sigma^{-1} s^{-1} \sigma s$, for σ^{-1} changes l_{hk} into h_k which s^{-1} changes into k which is changed by σ into l_k which finally s changes into h_{lk}. But that $\psi_\rho = \psi$ is manifest since by hypothesis $\psi_\sigma = \psi$ and hence $\psi_{\sigma^{-1}} = \psi$ also, but then $\psi_{\sigma^{-1} s^{-1}} = \psi_{s^{-1}}$ and $\psi_{\sigma^{-1} s^{-1} \sigma} = \psi_{s^{-1} \sigma} = \psi_{s^{-1}}$, hence finally $\psi_{\sigma^{-1} s^{-1} \sigma s} = \psi_{s^{-1} s} = \psi$. Thus if ψ remain unaltered for σ it also remains unaltered for ρ; which enables us to show that ψ is unchanged for at least one circular substitution of three elements. For whatever σ may be it is of one of the five following types :

1° σ contains at least one cycle of more than three letters and is of the form

$$\sigma_1 = \begin{pmatrix} 1\,2\,3\cdots m \\ 2\,3\,4\cdots 1 \end{pmatrix} c_1\, c_2 \cdots\cdots$$

where $c_1\, c_2 \cdots$ denote other cycles of σ_1

2°   it contains two or more cycles of three elements, but no cycle of more elements, and

$$\sigma_2 = \begin{pmatrix} 1\,2\,3 \\ 2\,3\,1 \end{pmatrix} \begin{pmatrix} 4\,5\,6 \\ 5\,6\,4 \end{pmatrix} c_1\, c_2 \cdots$$

3°   it contains only one cycle of three elements and one or more transpositions

$$\sigma_3 = \begin{pmatrix} 1\,2\,3 \\ 2\,3\,1 \end{pmatrix} \begin{pmatrix} 4\,5 \\ 5\,4 \end{pmatrix} c_1\, c_2 \cdots$$

4°   it contains only transpositions, but of these at least three

$$\sigma_4 = (1\,2)\,(3\,4)\,(5\,6)\, c_1\, c_2 \cdots$$

5°   it contains only two transpositions

$$\sigma_5 = (1\,2)\,(3\,4)\,(5)\cdots$$

Put now $\rho_\lambda = \sigma_\lambda^{-1} \sigma_\lambda^{-1} \sigma_\lambda s_\lambda$

$$s_1 = \begin{pmatrix} 1\,2\,3 \\ 2\,3\,1 \end{pmatrix} \quad s_2 = \begin{pmatrix} 1\,4\,5 \\ 5\,4\,1 \end{pmatrix} \quad s_3 = \begin{pmatrix} 1\,2\,4 \\ 2\,4\,1 \end{pmatrix} \quad s_4 = \begin{pmatrix} 1\,3\,5 \\ 3\,5\,1 \end{pmatrix} = s_5$$

then

$$\rho_1 = \begin{pmatrix} 1\,2\,4 \\ 2\,4\,1 \end{pmatrix} \quad \rho_2 = \begin{pmatrix} 1\,4\,5\,2\,6 \\ 4\,5\,2\,6\,1 \end{pmatrix} \quad \rho_3 = \begin{pmatrix} 1\,2\,5\,3\,4 \\ 2\,5\,3\,4\,1 \end{pmatrix}$$

$$\rho_4 = \begin{pmatrix} 1\,3\,5 \\ 3\,5\,1 \end{pmatrix} \begin{pmatrix} 2\,6\,4 \\ 6\,4\,2 \end{pmatrix}, \quad \rho_5 = \begin{pmatrix} 1\,3\,5\,4\,2 \\ 3\,5\,4\,2\,1 \end{pmatrix}$$

which shows that all the cases may be reduced to 1°. Hence ψ remains unaltered for some circular substitution of three elements

$$\tau = \begin{pmatrix} 1 & 2 & 3 \\ 2 & 3 & 1 \end{pmatrix}$$

But if ψ remain unchanged for τ it will also remain invariant for

$$(5) \quad \begin{pmatrix} 1 & 2 & 3 \\ 2 & 3 & 1 \end{pmatrix}, \quad \begin{pmatrix} 1 & 2 & 4 \\ 2 & 4 & 1 \end{pmatrix}, \quad \begin{pmatrix} 1 & 2 & 5 \\ 2 & 5 & 1 \end{pmatrix}, \quad \cdots \begin{pmatrix} 1 & 2 & n \\ n & 2 & 1 \end{pmatrix}$$

For let $t = \begin{pmatrix} 1 & 2 & \nu \\ 2 & \nu & 1 \end{pmatrix}$, then ψ being invariant for τ is for

$t^{-1} \tau t^{-1} = \omega$ and hence for ω; but $\omega = \begin{pmatrix} 2 & \nu & 3 \\ \nu & 3 & 2 \end{pmatrix}$ and hence

$\omega^2 = \begin{pmatrix} 2 & 3 & \nu \\ 3 & \nu & 2 \end{pmatrix}$ $\nu = 4, 5 \cdots n$

Every substitution of Γ is the product of the substitutions of (5), for every substitution of Γ is the product of a certain number of pairs of transpositions

$$(1\ 2) \quad (1\ 3) \quad \cdots (1\ n).$$

Let $(1\lambda)\ (1\mu)$ be such a pair, then as

$$(1\lambda)\ (1\mu) = \begin{pmatrix} 1 & 2 & \lambda \\ 2 & \lambda & 1 \end{pmatrix} \begin{pmatrix} 1 & 2 & \lambda \\ 2 & \lambda & 1 \end{pmatrix} \begin{pmatrix} 1 & 2 & \mu \\ \mu & 2 & 1 \end{pmatrix}$$

ψ remains unaltered for every substitution of Γ.

This demonstration, while longer than either of the two preceding ones, is interesting from a historical view, as the lemma on the conjugate functions ψ contains as a corollary Cauchy's theorem, which is fundamental in Abel's proof. In fact let

$$\varphi_1 \quad \varphi_2 \quad \cdots \varphi_m$$

be the m distinct values which a rational function φ of $x_1 x_2 \cdots x_n$ assumes for the substitutions of Γ; as no substitution of Γ leaves *all* the φ's unchanged every substitution of Γ will give rise to a permutation of the above m indices; as there are only $m!$ permutation of m things this number must be certainly as great as $\frac{1}{2} n!$ and

$$m! \gtreqless \tfrac{1}{2} n!$$

Now if $m < n$ then not only is $m! < n!$ but also $m! < \frac{1}{2} n!$ which is impossible. Thus every rational function φ of $x_1 x_2 \cdots x_n$ which takes on more than two values, takes on at least n values; which is Cauchy's theorem.

§ 7.

It being demonstrated that the general equation of degree $n > 4$ possesses no algebraic solution when the coefficients are regarded as independent variables, it lies very near to inquire whether there are equations with constant coefficients, say with rational-number coefficients, which do not possess an algebraic solution. To this very important question the Ruffini-Abelian theorem gives no reply; however in looking over the reasoning employed in the last three paragraphs we observe that its correctness depends upon the obvious fact that every rational equation

$$(1) \qquad \Phi\,(x_1\,x_2 \cdots x_n\,a,\,\beta \cdots) = 0$$

between the roots and certain roots of unity $a, \beta \cdots$, the coefficients being rational numbers, remains true on applying to it any substitution of the alternate group Γ. The question is thus reduced to this: do there exist equations of every degree, whose coefficients are rational numbers such that every relation (1) still subsists on applying any substitution of Γ? That this is so has been shown by Hilbert.* Hence not only has the *general* equation of degree > 4 no algebraic solution, but there exists an infinity of equations of every degree with integral coefficients which possess this property.

§ 8.

We close by giving an illustration of §§ 2, 3 applied to the explicit algebraic function

$$x = \sqrt[3]{-\frac{q}{2} + \sqrt{\frac{q^2}{4} + \frac{p^3}{27}}} + \sqrt[3]{-\frac{q}{2} - \sqrt{\frac{q^2}{4} + \frac{p^3}{27}}}$$

root of the cubic

$$(1) \qquad\qquad x^3 + px + q = 0$$

Here

$$R_1^2 = f_1(p,q) = \frac{q^2}{4} + \frac{p^3}{27}; \; R_2^3 = f_2(p,q,R_1) = -\frac{q}{2} + R_1$$

$$(a) \qquad R_3^3 = f_3\,(p,q,R_1,R_2) = -\frac{q}{2} + R_1$$

$$(2) \qquad\qquad x_0 = R_2 + R_3$$

* CRELLE, vol. 110, p. 104.

We proceed to replace the *suite* (α) by a *suite* (β) of the type (B) of §3. Raising (2) to the 2nd and 3rd powers and substituting in (1) we get

$$R_3^3 + 3\, R_2\, R_3^2 + (3\, R_2^2 + p)\, R_3 + (R_2^3 + p\, R_2 + q) = 0$$

or reducing by aid of

$$R_3^3 = f_3 = -\frac{q}{2} - R_1 \qquad R_2^3 = -\frac{q}{2} + R_2$$

we get

$$3\, R_2 . R_3^2 + (3\, R_2^2 + p)\, R_3 + p\, R_2 = 0$$

As the coefficients of this equation are not zero, the last equation of (α) cannot be reducible and R_3 must be rational in the preceding radicals. In fact we know that

$$R_3 = -\frac{p}{3\, R_2}$$

Equation (2) becomes thus

(3)
$$x_0 = R_2 - \frac{1/_3\, p}{R_1 - \frac{1}{2}\, q}\, R_2^2$$

The denominator here is $-R_1 - \frac{1}{2} q$ whose conjugate value is $-R_1 - \frac{1}{2} q$; hence multiplying numerator and denominator of the second term of the right member of (3) by this last, (3) becomes

(4)
$$x_0 = R_2 - \frac{9}{p^2}(R_1 + \tfrac{1}{2} q)\, R_2^2$$

which is an integral function of R_1, R_2 with coefficients rational in p, q. The value of x_0 given by (4) substituted in (1) gives

$$0 = B_0 + B_1\, R_2 + B_2\, R_2^2,$$

where

$$B_0 = R_1 - \frac{q}{2} - \frac{3^6}{p^6} \cdot \frac{p^6}{3^6}(R_1 + \tfrac{1}{2} q)$$

$$B_1 = -\frac{3^3}{p^2}\frac{p^3}{3^3} + p$$

$$B_2 = \frac{3^5}{p^4}\frac{p^3}{3^3}(R_1 + \tfrac{1}{2} q) - \frac{3^2}{p}\left(R_1 + \frac{q}{2}\right)$$

which are all identically equal to zero. Hence not only is (4) a root of (1) but so also are the quantities which we deduce from it in replacing R_2 by ωR_2 and $\omega^2 R_2$ $(\omega^3 = 1)$. Denoting these roots respectively by x_0 x_1 x_2 we have

$$x_0 = R_2 - \frac{3^2}{p^2}(R_1 + \tfrac{1}{2}q) R_2^2$$

$$x_1 = \omega R_2 - \frac{3^3}{p^2}(R_1 + \tfrac{1}{2}q)\omega^2 R_2^2$$

$$x_2 = \omega^2 R_2 - \frac{3^2}{p^2}(R_1 + \tfrac{1}{2}q)\omega R_2^2$$

As the coefficient of R_2 is here 1 we do not need to replace R_2 by a new radical. Solving we get

$$R_2 = \tfrac{1}{3}\sum \omega^{-k} x_k \qquad k = 0, 1, 2$$
$$= \frac{x_0 + \omega^2 x_1 + \omega x_2}{3}$$

The suite (α) is now replaced by

(β)
$$R_1^2 = f_1 \quad R_2^3 = f_2$$
$$x_0 = g(p, q, R_1, R_2,)$$
$$= R_2 - \frac{9}{p^2}(R_1 + \tfrac{1}{2}q) R_2^2$$

and the first step has been accomplished.

We have now to reason in the same way in regard to the remaining radical R_1. To this end we form the equation $\varphi(\xi) = 0$ whose roots are the values of

$$\xi_0 = R_2^3 = \left(\frac{x_0 + \omega^2 x_1 + \omega x_2}{3}\right)^3.$$

As ξ_0 acquires only two distinct values, φ will be the cube of an integral function of ξ with coefficients rational in p, q; this is easily found to be

$$\varphi(\xi) = \xi^2 + q\xi - \tfrac{1}{27} p^3,$$

Substituting here the value of ξ given in (β), $\varphi(\xi) = 0$ becomes

$$(R_1 - \tfrac{q}{2})^2 + q(R_1 - \tfrac{q}{2}) - \tfrac{1}{27} p^3 = 0,$$

or

$$B_0 + B_1 R_1 = 0$$

where

$$B_0 = \tfrac{1}{4} q^2 + \tfrac{1}{4} q^2 - \tfrac{1}{2} q^2 + \tfrac{1}{27} p^3 - \tfrac{1}{27} p^3$$
$$B_1 = q - q$$

vanish identically; whence not only

$$\xi_0 = R_1 - \frac{q}{2}$$

but also

$$\xi_1 = -R_1 - \frac{q}{2}$$

is a root of $\varphi = 0$.

Whence

$$R_1 = \tfrac{1}{2}(\xi_0 - \xi_1) = \tfrac{1}{2 \cdot 3^3} \left[(x_0 + \omega^2 x_1 + \omega^2 x_2)^3 - (x_0 + \omega x_1 + \omega^2 x_2)^3 \right]$$

$$= -\frac{\sqrt{-3}}{2.9} \sqrt{\triangle},$$

where $\triangle = (x_0 - x_1)^2 (x_1 - x_2)^2 (x_1 - x_2)^2$ is the discriminant of (1). Thus the *suite* (β) has the character (B).

YALE UNIVERSITY,
  NEW HAVEN, CONN.

EARLY HISTORY OF GALOIS' THEORY OF EQUATIONS.

BY PROFESSOR JAMES PIERPONT.

(Read before the American Mathematical Society at the Meeting of February 26, 1897.)

THE first part of this paper will treat of Galois' relations to Lagrange, the second part will sketch the manner in which Galois' theory of equations became public. The last subject being intimately connected with Galois' life will afford me an opportunity to give some details of his tragic destiny which in more than one respect reminds one of the almost equally unhappy lot of Abel.* Indeed, both Galois and Abel from their earliest youth were strongly attracted toward algebraical theories ; both believed for a time that they had solved the celebrated equation of fifth degree which for more than two centuries had baffled the efforts of the first mathematicians of the age; both, discovering their mistake, succeeded independently and unknown to each other in showing that a solution by radicals was impossible ; both, hoping to gain the recognition of the Paris Academy of Sciences, presented epoch making memoirs, one of which

---

* Sources for Galois' Life are :

1° Revue Encyclopédique, Paris (1832), vol. 55.

α Travaux Mathématiques d Évariste Galois, p. 566-576. It contains Galois' letter to Chevalier, with a short introduction by one of the editors.

β Ibid. Nécrologie, Évariste Galois, p. 744-754, by Chevalier. This touching and sympathetic sketch every one should read.

2° Magasin Pittoresque. Paris (1848), vol. 16. Évariste Galois, p. 227-28. Supposed to be written by an old school comrade, M. Flaugergues, an ardent admirer of Galois.

3° Mes Mémoires, A. Dumas, (père). Paris (1869), vol. 8, p. 159-161 and p. 166-169. Gives an account of the notorious affair of the Vendanges de Bourgogne and of Galois' first trial.

4° Journal de Mathématiques, (1846), v. 11. Oeuvres Mathématiques d'Évariste Galois, p. 381-444. The avertissement is by Liouville and contains some personal notes about Galois.

5° Nouvelles Annales de Mathématiques. Paris (1849), vol. 8, p. 452. A few lines only.

6° Annales de l'École Normale. Paris (1896), 3 Ser. vol. 13, p. 197-266. La Vie d'Évariste Galois, by P. Dupuy, professor of history in the École Normale. This is the most extensive biography of Galois. It contains all that is essential in the five preceding references and an immense amount of other matter. It goes without saying that I am indebted to it for many of the details here given.

7° Oeuvres Mathématiques d'Évariste Galois. Paris, 1897. . The introduction by Picard contains interesting remarks.

Reprinted from Bull. Amer. Math. Soc. **4** (April 1898), 332–340

was lost forever, the other published first twelve years after the death of its illustrious author ; in the case of both the passionate feverish blood of genius led them to a premature death, one at the age of twenty, the other at twenty-six. Both died leaving behind them theories whose thresholds they had hardly crossed, theories which, belonging to the first creations of this century, will keep their memories perennially green. I shall have no occasion in this paper to speak of anything except Galois' algebraical theories. It is, however, only just to his memory to recall a fact too universally overlooked that in spite of his extreme youth he was perfectly familiar with the researches of Abel and Jacobi and that he was certainly in possession of the most essential results of Riemann in regard to Abelian integrals discovered twenty-five years later. A remark he makes regarding a generalization of Legendre's equation connecting the periods of elliptic integrals of the first and second species

$$FE' + EF' - FF' = \frac{\pi}{2}$$

leads us to suspect that he had also anticipated some of the results of Weierstrass and Fuchs in this field. All goes to show the justness of Picard's remark * that "if only a few years more had been given him to develop his ideas in this direction he would have been the glorious continuator of Abel and would have erected in its most essential parts the theory of algebraic functions of one variable as we know it to-day."

1. *Galois' relations to Lagrange.*

It is well known that Galois, like Abel and Ruffini, received inspiration from the writings of Lagrange regarding the algebraic solution of equations, in particular his mémoire "Sur la résolution algébrique des équations" and his " Traité de la résolution des équations, etc.," but no one, I believe, has remarked how well Lagrange had prepared the way for Galois.

Let us capitulate rapidly the principal facts of Lagrange's theory referring for more detail to my paper† on "Lagrange's place in the theory of substitutions." In the first place he gave the theory of the solution of general equations $f(x) = 0$ of degree n an entirely new basis in employing rational func-

---

* Introduction to the new edition of Galois' Oeuvres, which has just appeared under the auspices of the Société Mathématique de France.

† BULLETIN, 2 Series, vol 1, p. 196. (1895.)

tions of the roots of f to build his resolvents. Two rational functions of the roots of f invariant for the same permutations of the roots he calls similar and shows that they are roots of rational equations whose degrees are equal and a divisor of n !; furthermore, they are rational in one another. If φ be a rational function of the roots taking on r values for all possible permutations and ψ another function which changes its value for all the permutations which alter φ as well as for some of the permutations which leave φ unaltered taking on in all rs values, then φ is rational in ψ while ψ is root of an equation of degree s whose coefficients are rational in φ. This being so, a function t which changes its value for every permutation enjoys the remarkable property that the roots themselves, $x_1, x_2, \cdots, x_n$, of f are rational in t. Such a function, and Lagrange remarks that its simplest form is

$$t = ax_1 + bx_2 + \cdots + lx_n,$$

is root of a rational equation of degree n !, and all its roots are rational in any one of them.

This resolvent plays an important part in Lagrange's theory; I shall call it Lagrange's resolvent and denote it by $L(t) = 0$. The solution of $f = 0$ and $L(t) = 0$ are evidently equivalent problems. Let us see how a general scheme for solving the equation $f = 0$ may be deduced from these facts. Take a rational function φ of the roots of f and form its resolvent. Its solution gives its roots $\varphi_1, \varphi_2, \varphi_3, \cdots$ as known algebraic functions of the coefficients of f. That is, certain rational functions of the roots of f are now known which at the start were not. The effect of this is to make the Lagrangian resolvent $L = 0$ which before was irreducible split up into a certain number of equal factors whose coefficients are rationally known. In other words, the determination of t which before depended upon an equation of degree n !, depends now upon an equation of less degree say $L_1(t) = 0$. Take now another rational function ψ of the roots of f which is not rational in the φ's. This will give rise to a resolvent whose coefficients are rational in the φ's ; its solution makes still other rational functions of the roots known which before were not and the effect of this is to make $L_1(t)$ reducible and thus t depends upon an equation of still less degree $L_2(t) = 0$ with known coefficients. Continuing in this way, the degree of the equation upon which t depends becomes less and less , at last it must depend upon an equation of first degree, and then t is rational in known quantities. But since the roots of f are rational in t, these roots are also rational in known quantities.

Now this is precisely the Galoisian scheme of solution. In other words, the task Galois had before him was: 1° to formulate the solution of *any* equation as here sketched and as indicated with considerable clearness in Lagrange's writings, and 2° to extend the theorems upon which this scheme of solution depends, from the limited case here considered, viz. equations whose coefficients are independent variables, to the case of any special equation and an arbitrary domain of rationality. The application Lagrange made of his theory to solve the special equations upon which the roots of unity of prime order depend was doubtless of great assistance to Galois in his work of generalization. Lagrange proves in this connection the cardinal fact that all cyclic functions of the roots are rationally known and uses this as the base of his solution. The cyclic group is precisely the group which here plays the same rôle as the symmetric group does for general equations.

2. *How Galois' algebraic theories became public.*

As already observed the manner in which this took place is so intimately connected with his short and agitated career that a few details of his life are indispensable.

Évariste Galois was born October 25, 1811, in the little town of Bourg-la-Reine, a few miles to the south of Paris. His father, Nicolas-Gabriel, as his grandfather before him, was proprietor of a flourishing school, established before the French Revolution and at Évariste's birth a corporate part of the University of France. His mother, Adélaïde Demante, was daughter and sister of eminent jurists of the Faculté de Droit at Paris. From her father she had received a careful and extensive education in the classics. Till Évariste was twelve years old she was his only teacher. At this stage, that is, in the autumn of 1823 he entered Louis-le-Grand, one of the celebrated lycées of Paris, a stone's throw from the Sorbonne and the Collége de France. At fifteen he first began the study of mathematics, entering the class known as *Mathématiques préparatoires*. A new world now opened up to him ; the study of mathematics became the absorbing passion of his life. The Geometry of Legendre he is said to have read through as one reads a novel. Soon ordinary treatises did not satisfy him ; they lacked, so he said, the stamp of the great inventors; he began the study of the original authors, especially Lagrange. In the Autumn of 1828 he entered the class of *Mathématiques spéciales*, under the charge of Professor

Richard, an excellent mathematician and a man of noble character.   Under his intelligent and sympathetic direction Galois' progress became even more remarkable.   His original solutions of problems proposed in the class were explained by Richard with just praise to his admiring classmates.   In the trimestrial reports of the school we find Richard writing : " Cet élève a une supériorité marquée sur tous ses condisciples," and again " Cet élève ne travaille qu' aux parties supérieures des Mathématiques," while reports on his other studies are a series of lamentations caused by Galois' neglect.   So, for example, this: " Il ne fait absoluement rien pour la classe.   C'est la fureur des Mathematiques qui le domine."   At this epoch, that is, when seventeen years old, Galois made his first discoveries in the theory of equations which can be solved by radicals.   He drew up a short mémoire on the subject and Cauchy took it to present to the Academy of Sciences.   The mémoire was never heard of again, although Galois reclaimed it several times at the secrétariat.   At this time also his first paper, entitled " Démonstration d'un théorème sur les fractions continues périodiques" was published, appearing in the March number of the *Annales de Gergonne* for 1829.

Galois ardently desired to enter the École Polytechnique, the first school of Mathematics in France.   Twice he presented himself for examination but was rejected both times to the astonishment of all who knew him.   Two reasons probably lead to this deplorable mistake.   Galois had the habit of working almost exclusively with his head and to deal with the broad aspects of a subject.   To his remarkable mind many things seemed trivial or self-evident, which required demonstration for those less gifted, and complaint had already been made on various occasions that he was somewhat obscure in the expression of his ideas.   When therefore at the examination it became necessary to work out on the blackboard, explaining as he proceeded, before a numerous audience questions of pure detail he was greatly embarrassed.   He did not have what one called *l' habitude du tableau*.   The second reason is that the examinators were flagrantly incapable of appreciating the extraordinary talents of the youth they had before them.   At the distance of twenty years the injustice of his examinators was still keenly felt, for Terquem in a short notice on Galois which appeared in the Nouvelles Annales* writes : " Nous répéterons ici et nous ne cesserons de répéter une réflexion que nous

---

* Vol. 8 (1849), p. 452.

avons déjà consignée mainte fois : Un candidat d'une intelligence supérieure est perdu chez un examinateur d'une
intelligence inférieure. *Barbarus hic ego sum quia non in-
telligor illis.* Certes M. Liouville qui nous a fait connaître
le génie de Galois ne l'aurait pas jugé incapable.''

Nothing was left to Galois but to enter the École Prépara-
toire,*a fate which filled his heart with bitterness and despair.
At this crisis Galois' father, whom he dearly loved, died a
violent death. As mayor of his native town, M. Galois
had defended for many years the interest of the liberals
against the grasping power of the clerical party. Driven at
last to desperation by their incessant and venomous attacks,
his kindly nature succumbed and in a moment of despon-
dency he committed suicide. Still Galois continued with
his mathematics, adding to his already vast stores of erudi-
tion and maturing his own inventions. During the first year
at the École, that is, when eighteen years old, he published
four mémoires. One of these is a brief summary of some of
his results regarding the algebraic solution of equations ;
another is devoted to the theory of those numbers which
we now call Galois' imaginaries and which play an important part in the theory of groups to-day. In January of
this year (1830) Galois presented to the Academy another
mémoire containing an account of his researches written out
with care and detail. He placed great hope in this. The
manuscript was given to Fourier to read, who shortly after
died, and the mémoire was lost. Six months later the Revolution broke out which drove Charles X. from the throne and
installed Louis Philippe in his stead. In this movement
Galois entered with heart and soul. The attitude which
the director of the school, M. Guigniault, took toward the
new government offended Galois so deeply that he was
rashly led to take part in a polemic which the *Gazette des
Écoles* was waging against Guigniault. The scandal that
this raised produced Galois' expulsion (Dec. 9, 1830).
Powerful friends however interested themselves in him. A
minister of the Royal Council, M. Barthe, summoned the
unfortunate youth and bade him not to be concerned for the
future. Poisson invited him to make a new redaction of
his researches and volunteered to present it himself to the
Academy. Inspired by fresh hopes Galois wrote the only
finished mémoire we have of his grand theory of the solution of equations. It bears the title '' Sur les conditions de

---

*This took the place of the École Normale suppressed in 1822.
Dupuy speaks of it as a faint and humble copy of this celebrated school.
After the revolution of 1830 it became the École Normale again.

résolubilité des équations par radicaux." At the same time he opened a course on Higher Algebra which was held in a bookstore near the Sorbonne. The *Gazette des Écoles* announced it as follows : "This course will be held Tuesdays at quarter past one ; it is destined for those who, feeling how incomplete courses on algebra are, desire to penetrate more deeply into this science. Many of the theories to be given are entirely new, none have ever been developed in university lectures. We content ourselves to mention a new theory of imaginaries, a theory of equations which are soluble by radicals, the theory of numbers, and the theory of elliptic functions treated entirely by algebra." But his hopes were short lived ; the mémoire intrusted to Poisson was returned, being declared unintelligible.

Already associated with the most reactionary faction of the republican party, Galois now plunged deeper and deeper into the political turmoil which was shaking France at its foundations. Twice arrested for political offenses, he spent the greater part of the last year and a half of his life in the prison of Sainte Pélagie. On being released the second time he became engaged in an affair of honor which resulted in his death (May 31, 1832). The night before the duel he spent in drawing up with feverish haste an account of his discoveries. No one can read this scientific testament written in the form of a letter to his friend and schoolmate, Auguste Chevalier, without emotion ; it is one of the most touching documents in the history of science. Cut off in the very commencement of his career, conscious of the importance of his discoveries, he spent the last few hours of his life trying to save to posterity what his contemporaries had been loth to accept. He requests Chevalier to publicly beg Gauss and Jacobi to give their opinion, not on the truth but upon the importance of his theorems. "Après cela," he adds, "il y aura, j'espère, des gens qui trouveront leur profit à déchiffrer tout ce gâchis."

Chevalier was thus appointed Galois' scientific executor. Conformably to Galois' wishes the letter we have been speaking of was published the following September in the Revue Encyclopédique and an editorial note states that all Galois' manuscripts would shortly appear in the same journal under the editorship of Chevalier. This was not to be the case. Nothing more was published until 1846 when a part of them, edited by Liouville, appeared in Vol. XI. of the Journal de Mathématiques. In the preface which accompanies this collection Liouville remarks : "When, at the desire of Évariste's friends, I began an attentive study of all his published mémoires and the manuscript he left behind, I considered it my sole task to disentangle, as far as possible, what there was new in these productions in order to make its value more evident. My zeal was soon rewarded and it gave me the greatest pleasure when, on filling some slight gaps, I recognized the entire correctness of the method whereby Galois proves this beautiful theorem : In order that an irreducible equation of prime degree be soluble by radicals it is necessary and sufficient that all its roots are rational in any two of them. This method, eminently worthy of the attention of geometers, will alone suffice to assure our compatriot a place among the few mathematicians who merit the title of inventor." Liouville's purpose was to publish all Galois' works and to add a commentary in which he intended to complete certain passages and elucidate various delicate points. Unfortunately this plan was never executed, only two new papers were published, the mémoire written at Poisson's invitation already mentioned and a fragment of a second mémoire entitled. " Des équations primitives qui sont solubles par radicaux." A note at the end of the preface informs us that the press of matter for publication as well as the extent of Galois' manuscript makes it necessary to give only a part in the current number, the rest following in the next volume. As we just said, this promise was not fulfilled.

The next reference I find to Galois' theory for the solution of algebraic equations is in a footnote, p. 344 of the first edition of Serret's Algèbre Supérieure, published in 1849 : " My friend, M. Liouville," he remarks, " has announced to me his intention to publish one day some developments regarding this remarkable work. It is only by aid of these developments, a part of which M. Liouville has kindly communicated to me, that I have succeeded in comprehending certain points of Galois' Mémoire, whose study only those geometers can undertake who have occupied themselves quite specially with the theory of equations. For these reasons I have been prevented from presenting here the discoveries of Galois." The cause of Liouville's silence can only be conjectured. It seems probable that, on a more careful revision, preparatory to publishing, certain points in his demonstrations did not seem to be rigorously established. At any rate, Betti, having asked him in a paper published in the Annali of Tortolini, in 1851, not to deprive the public any longer of the results of his study, proceeded to publish his own commentary in the same journal the following year. Thus in 1852, twenty years after Galois' death, his theory of the resolution of algebraic equations was for the first time made intelligible to the general public and established with complete rigor. A few words more on the further history of Galois' theory will complete my account. According to Weber,* Kronecker probably first became acquainted with it during his visit to Paris in 1853 where he was associated intimately with Hermite, Bertrand, and other leading French mathematicians. The first mention Kronecker makes of Galois' name is in a letter to Dirichlet in March, 1856. Dedekind also became very early acquainted with Galois' theory since it is known that he lectured in the winter of 1857–58 on higher algebra and in particular on Galois' theory. According to Weber† this was probably the first extensive account of Galois' theory given at a German university. The first account of it given in a text-book on algebra is in the third edition of Serret's algebra (1866). This, together with Jordan's classic treatise which appeared in 1870, made a knowledge of Galois' theory possible to all the world.

Perfectly just was Galois' estimate of his own discoveries when he said shortly before his death: "*J'ai fait des recherches qui arrêteront bien des savants dans les leurs.*"

YALE UNIVERSITY,
February, 1898.

Chapter 4

Henri Poincaré

Henri Poincaré clearly was a dominating force in our field, A succinct review of his family is given in a July 17, 1912, *The Times* of London obituary of his unexpected and sudden death (embolism of the heart as he was leaving after recovering from an operation).

> *"The death of Henri Poincaré at the comparatively early age of 58 deprives the world of one of its most eminent mathematicians and thinkers. ... He was born at Nancy in 1854, and was a member of an old* bourgeois *family. His father was a Lorraine doctor of some eminence, and his grandfather had been a chemist. His uncle, an inspector of roads and bridges, was father of another eminent Academician, M. Raymond Poincaré, now Prime Minister* [and later he became President] *of France and Minister for Foreign Affairs. A sister of Henri Poincaré is the wife of M. Emile Boutroux, the well-known writer on moral philosophy, member of the Institute, and Director of the Thiers Foundation for the Promotion of Higher Learning."*

The Times goes on to briefly describe his early life and career. In this description notice that at the tender age of 16, Poincaré was involved, in a non-combat role, in the French-German war of 1870. Another interesting comment, as suggested by the article, is how the lure of mathematics pulled Poincaré away from a career of mining.

> *"Henri Poincaré began his education at the Lycée of Nancy. During the war of 1870 he served, as a mere boy, in the Ambulance Corps with his father. He passed with great distinction through the Ecole Polytechnique, and intended to devote himself to the*

scientific study of mining. He made several expeditions as a min-
ing school pupil to Austria, and actually obtained an appointment
as mining engineer as Vesoul. But in 1879 he had taken his de-
gree as Docteur lés Sciences, and he shortly afterwards accepted
an appointment as lecturer at Caen. His work was so brilliant
that he was transferred to the University of Paris in 1881, where
he lectured first on physcial mechanics, then on mathematical
physics and the calculation of probabilities, and ultimately on
astronomical mechanics."

As a slight digression, wars affect all, so it is not surprising that the 1870 French-German conflict affected other mathematicians. One story, which involves two mathematicians represented elsewhere in this volume, is somewhat indicative of our field. This is where the Norwegian mathematician Sophus Lie, (1842-1899) left Paris at the outbreak of the French-German war for the safer setting of Italy. En route, however, the French arrested him as a German spy. The "evidence" involved his papers which were presumably written in some secret code. As Stubhaug [14] writes, "When Lie asked a guard what was done with prisoners like himself, the reply was as follows: 'We usually shoot them at six in the morning.' " After spending a month in a prison at Fontainebleau, Lie was released after the French mathematician Gaston Darboux, one of the few mathematicians who could read Lie's work and had been promoting his discoveries, finally convinced the authorities that this "secret code" was just Lie's mathematical notes.

A large portion of Poincaré's work centered on, or was motivated by, celestial mechanics. And what a tremendous, lasting contribution! An often told story within the dynamical systems community describes how the birth of chaos theory was a direct consequence of Poincaré's efforts to explain and understand why his earlier attempt to solve the three-body problem (actually, the restricted three-body problem) was in error. In recent years, this story has been made available to the general mathematical community with readable books by Barrow-Green [4] and by Diacu and Holmes [5].

A consequence of Poincaré's efforts, which describes several of the innovative approaches he introduced to this area, was his three volume series *Méthodes nouvelles de la Mécanique Céleste*. This work, full of mathematical nuggets, continues to be read today; e.g., in the late 1980s and early 1990s, mathematicians and mathematical astronomers in Paris interested in reviewing this series gathered at Bureau des Longitudes to lecture on these books; there are reports [1, 2, 6] of their findings.

An early description of Poincaré's series *Méthodes nouvelles de la Mécanique Céleste* is the first article reproduced in this section; it appeared in Ernest W. Brown's [3] *Bulletin* review. Brown (1866-1938), who served as the AMS President during 1915-16, was in the mathematics departments of Haverford College in Pennsylvania from 1891-1907 and Yale University (1907-1932). His research emphasized the more practical aspects of celestial mechanics concerning the dynamics and motion of the moon. (For more about Brown, see Schlesinger and Brouwer [13].)

As the *Times* obituary noted, Poincaré was one of the world's most eminent thinkers. A taste of this is found in the second article reproduced in this section, "The relations of analysis and mathematical physics;" this is the written version of Poincaré's address [8] at the 1897 International Congress of Mathematicians held in Zurich. Poincaré starts his informal address in an unusual, enjoyable manner.

> *"Doubtless you are often asked what is the utility of mathematics and whether its nicely constructed theories, drawn entirely from the mind, are not artificial products of our own caprice.*
>
> *Among the persons who ask this question, I must make a distinction. The practical class demand of us nothing but means of getting money. These do not deserve to be answered. Rather ought it to be demanded of them what is the good of accumulating so much wealth and whether, in order to have time for its acquisition, it is necessary to neglect art and science, which alone render the soul capable of enjoying it."*

Poincaré goes on to show "... precisely the nature of the relations between pure science and its applications." He argues, and demonstrates, that "Mathematical physics and pure analysis are not simply adjacent powers maintaining the relations of good neighborhood; they interpenetrate, and their spirit is the same." Beyond its own interest, in a later chapter I describe how this paper has a loose connection with Hilbert's list of problems.

As recently as 2000, I was at a conference where an expert in mathematical physics argued how Poincaré, not Einstein, should be treated as the father of special relativity. This ongoing debate has persisted almost since the publication of Einstein's contributions. Much of the argument centers around a paper Poincaré delivered at the St. Louis exposition that was described earlier; the first English translation of Poincaré's address was published in the *Bulletin* [9]. In his presentation, Poincaré appraised the challenges faced by twentieth-century mathematical physics. In part, he describes the then hot topic of *relativity* ending with his challenge that perhaps "we shall have to construct an entirely new mechanics;" a challenge that was met the following year by A. Einstein.

Everyone agrees that the contributions of Poincaré and Lorentz must be recognized. but, how much credit should Poincaré be given? Poincaré's paper is republished here because it provides an excellent insight into the mathematics of the day. But, the paper is available, so you are interested in the relativity debate, read it and form your own opinion.

Although Poincaré was one of the more gifted expositors in our field, it appears that this was not always the case. For instance, it is amusing to learn from Miller [7] the examiners' reaction to his 1879 thesis where they noted that

> "... the thesis is a little confused and shows that the author was still unable to express his ideas in a clear and simple manner."

Poincaré played an important role in the crucial concern of presenting the meaning and importance of science and mathematics to the general public. His more popular books include *Science and Hypothesis* [10], *The Value of Science* [11], and *Science and Method* [12]. The last article reproduced in this section is Wilson's [15] review of the first of these books. As Wilson starts,

> "Not logical enough for the logician, not mathematical enough for the mathematician, not physical enough for the physicist, not psychological enough for the psychologist, nor metaphysical enough for the metaphysician, Poincaré's Science and Hypothesis can hardly give the satisfaction of finality to any one; and yet it probably comes nearer to satisfying the requirements of all these classes of investigators than any single book of our acquaintance."

High praise.

Bibliography

[1] Chenciner, A., J. Laskar, Méthodes nouvelles de la Mécanique Céleste (H. Poincaré). Volume 1, Notes techniques du Bureau des Longitudes S026, Mai 1989.

[2] Chenciner, A., Méthodes nouvelles de la Mécanique Céleste. Séries de Lindstedt. Volume 2. Notes techniques du Bureau des Longitudes S028. Mars 1990.

[3] Brown, E. W., Book review of "Poincaré's Mecanique Celeste" *BAMS* 206-214.

[4] Barrow-Green, J., *Poincaré and the Three Body Problem*, AMS, London Math Society, 1997,

[5] Diacu, F., and P. Holmes, *Celestial Encounters; The Origins of Chaos and Stability*, Princeton University Press, Princeton, 1996.

[6] Ferraz-Mello, S., A. Jupp, Méthodes nouvelles de la Mcanique Céleste (H. Poincaré). Volume 3. Notes techniques du Bureau des Longitudes S036, Mai 1992.

[7] Miller, A., *Insights of genius : Imagery and creativity in science and art*, MIT Press, 2000.

[8] Poincaré, H., The relations of analysis and mathematical physics (An address before the International Congress of Mathematicians, Zurich, August 9, 1897), *BAMS* 4, 247-255.

[9] Poincaré, H., The present and the future of mathematical physics, *BAMS* **12** (1906), 240-260.

[10] Poincaré, H., *Science and Hypothesis* (1901), reprinted, Dover 1952.

[11] Poincaré, H., *The Value of Science* (1905), reprinted, Dover, 1958.

[12] Poincaré, H., *Science and Method* (1908), reprinted, Dover, 1958.

[13] Schlesinger, F. and D. Brouser, *Biographical Memoirs of the National Academy of Science* **21** (1941), 243-73.

[14] Stubhaug, A., *The Mathematician Sophus Lie; It was the Audacity of My Thinking*, Springer, 2002.

[15] Wilson, E., The foundations of science, *BAMS* **11** (1906), 187-193.

POINCARÉ'S MÉCANIQUE CÉLESTE.

Les Méthodes nouvelles de la Mécanique Céleste. Par H. POINCARÉ. Tome I. Paris, Gauthier-Villars, 1892. 8vo.

THE publication of this new work on Celestial Mechanics, embodying some of the results of the labors of mathematicians in that direction during the last fifteen years, comes as a welcome addition to our knowledge of this subject. Until lately, nearly all treatises have been written with a special object, that of obtaining expressions which can be used by the practical astronomer ; the mathematical aspects of the problems solved have been almost entirely neglected. These latter have an interest of their own apart from any use which can be made of them, and it is to the study of such questions that M. Poincaré largely devotes himself. At the same time he points out where they can be applied usefully in the case of the problem of three bodies. But this is not all. Most of the results obtained can be applied equally to the general problems of dynamics where there is a force function, and by the use of a dissipation function could doubtless be applied to any natural problem whatever.

The applications are, however, more particularly made to a satellite system, in the special case when the three bodies move in one plane, as well as in the general case. The limitation generally imposed consists in making the ratios of the masses of two of the bodies to that of the third a small quantity, an assumption which, nevertheless, does not limit greatly the usefulness of the results. M. Poincaré says, " Le but final de la Mécanique céleste est de résoudre cette grande question de savoir si la loi de Newton explique à elle seule tous les phénomènes astronomiques," and for this end to be attained it is absolutely necessary to know whether the developments of the expressions for the position of any heavenly body do mathematically represent that position. In general, the series obtained must be convergent, and it is to the questions on the convergence of such series that M. Poincaré has been able to give some definite answers.

In his introduction, the author points out that the starting point of the present developments of the lunar theory, was the publication in Vol. I. of the *American Journal of Mathematics* of a paper by Dr. Hill entitled, " Researches in the lunar theory." It is true that in this memoir, Dr. Hill has largely occupied himself in obtaining exact numerical and algebraical values for certain inequalities in the motion of the moon ; but the general considerations involved at the beginning and end of it are of a far-reaching nature. In par-

Reprinted from Bull. New York Math. Soc. 1 (1891–92), 206–214

ticular, a superior limit to the radius vector of the moon is
found, and a general study of the surfaces of equal velocity is
made. His consideration in a particular case of the moons
of different lunations with respect to the primary, will be mentioned below.

M. Poincaré's book is principally based on his own memoir,
"*Sur le problème des trois corps et les équations de la
dynamique.*"* The arrangement is not quite the same. In
the treatise, many of the demonstrations are more completely
explained, the applications are more numerous, and much
matter that is entirely new has been added. In what follows,
I have not in any sense attempted to give a complete account
of the book. Much that is given there is outside the scope
of an article such as this; the results that are mentioned
are chiefly noticed either because they can be given in a few
words, or because from their peculiar interest they merit a
somewhat longer treatment.

The first chapter deals with some general well-known theorems with respect to differential equations. Two types are
selected. The general form which it is necessary to consider
is shown by the system

$$\frac{dx_i}{dt} = X_i \qquad (i = 1, 2, \ldots n).$$

The X_i are analytic and single-valued functions of the x_i and
may or may not contain the time explicitly. This type includes the system of canonical equations

$$\frac{dx_i}{dt} = \frac{dF}{dy_i}, \qquad \frac{dy_i}{dt} = -\frac{dF}{dx_i},$$

which possess a set of properties special to themselves. Some
space is devoted to the consideration of these properties, and
special attention is directed to changes of variables for
which the system still remains canonical. The proofs for
these theorems are sketched very briefly in cases where they
are well-known.

In all the particular cases of the applications of canonical
equations to the problem of three bodies, M. Poincaré works
out the results with some detail. The masses are taken to be
m_1, m_2, m_3; m_1 is the mass of the primary while m_2, m_3 satisfy

$$\beta\mu = \frac{m_1 m_2}{m_1 + m_2}, \qquad \beta'\mu = \frac{(m_1 + m_2) m_3}{m_1 + m_2 + m_3},$$

---

* *Acta Mathematica*, Vol. XIII.

such that μ is small while β, β' remain finite quantities. It is then possible to expand F in a series arranged in ascending powers of μ :

$$F = F_0 + \mu F_1 + \mu^2 F_2 + \cdots$$

In general F_0 will be independent of one system of canonical elements, say, the y_i.

The canonical equations given above correspond to n degrees of liberty. If we know an integral of the system, this number can be lowered by one unit. In general, if we know q integrals, Poisson's conditions must be fulfilled between these integrals taken two and two, in order that the number of degrees of liberty may be lowered by q units. The application of this to the general problem of three bodies is immediate. The three integrals for the motion of the centre of mass of the system being known and fulfilling the conditions, we can reduce the number of degrees of liberty from *nine* to *six*. The three known integrals of areas are also integrals of the system thus reduced, and by using two combinations of these latter, it is possible to reduce the system to *four* degrees of liberty ; also in the case when the bodies move in one plane, the system can be reduced to *three* degrees of liberty. The usual transformations are then effected so as to leave the equations still in the canonical form and to carry only the smallest number of degrees of liberty.

The form of the disturbing function is also discussed, and it is considered under what circumstances we can develop it in ascending powers.

The second chapter deals with the general conditions for integration in series, and in particular with the conditions that these series may be convergent. It is here that M. Poincaré's penetrative genius especially shows itself. The complicated forms which appear in the lunar problem render it an almost impossible task to attack directly the question of convergence of the series obtained. But by going back to the differential equations themselves, and considering the disturbing function, he is able to obtain definite results, with respect to the problem of three bodies, for the convergence of those series which may be taken to represent certain particular solutions.

The notation introduced by M. Poincaré a short time back for dealing with questions of convergence is an especially happy one. It is as follows :—If we have two functions φ, ψ, expanded in ascending powers of x, y,

$$\varphi \ll \psi \qquad (\text{arg. } x, y)$$

denotes that the coefficient of every term in ψ is greater in absolute value than the corresponding term in φ, the " arguments " in terms of which the expansion is made being written as above. This can of course be used for any number of arguments. An extension of this notation is given at the end of the chapter. The coefficients, instead of being constants, are supposed to be periodic functions of the time ; then, if every coefficient of ψ in its expansion according to powers of x, y, $e^{\pm u}$ is real, positive, and greater in absolute value than the corresponding coefficient in φ,

$$\varphi \ll \psi \qquad (\text{arg. } x, y, e^{\pm u}).$$

Cauchy's general theorems on convergence are quoted and extended to the case in which the function is expanded in terms of several variables. If we have a system of differential equations

$$\frac{dx}{dt} = \theta(x, y, z, \mu), \frac{dy}{dt} = \varphi(x, y, z, \mu), \frac{dz}{dt} = \psi(x, y, z, \mu)$$

where θ, φ, ψ are expanded in powers of x_0, y_0, z_0, and μ, t, there will exist three series expanded in powers of x_0, y_0, z_0 and μ, t which will satisfy these equations and reduce respectively to x_0, y_0, z_0 when $t = 0$. For these to be convergent it is necessary that $|x_0|$, $|y_0|$, $|z_0|$, $|\mu|$, $|t|$ should be sufficiently small. The restriction $|t|$ sufficiently small is evidently inconvenient, and Poincaré is able to get rid of it and to say that the series are convergent if t lies between given limits provided that $|\mu|$ be sufficiently small.

In most cases, however, expansion is not made in powers of the time, but in trigonometrical functions of it, and it therefore becomes necessary in the first instance to examine a system of differential equations,

$$\frac{dx_i}{dt} = \varphi_{i,1}x_1 + \varphi_{i,2}x_2 + \ldots + \varphi_{i,n}x_n \ (i = 1, 2, \ldots n)$$

where the φ are all periodic functions of the time. The general solution found is,

$$x_i = c_1 e^{a_1 t} \lambda_{1,i} + c_2 e^{a_2 t} \lambda_{2,i} + \ldots + c_n e^{a_n t} \lambda_{n,i}$$

the λ being periodic functions of the time only, the a_i dependent on the roots of a determinantal equation, and the c_i arbitrary constants.

These a_i are called the characteristic exponents (*exposants caractéristiques*) of the solution. On them depends the whole nature of the various solutions. Thus if two of the exponents are equal, the time appears as a factor ; if they are all

14

pure imaginaries, the general solution contains periodic terms
only, and so on. Also, on them depend the " asymptotic solutions." Chapter IV. is devoted to the consideration of these
exponents.

Chapter III., which deals with periodic solutions, is perhaps
the most interesting from the point of view of its immediate
application to some of the problems in the lunar theory. In
this connection, a periodic solution is defined as being such
that the system at the end of a finite time T comes into the
same *relative* position as at the beginning of that time. The
period is then T. Thus if $\varphi(t)$ represent a periodic solution
of period T

$$\varphi(t + T) = \varphi(t) ;$$

also if

$$\varphi(t + T) = \varphi(t) + 2k\pi \qquad (k = \text{whole number})$$

$\varphi(t)$ is still said to be a periodic solution. These two types
are analogous to linear and angular coördinates, respectively.
In the canonical system of coördinates as applied to dynamical
problems, one set of elements belongs to the first type, and the
conjugate set in general to the second type. It is to be noted
that by defining a periodic solution in this way, the system
can, so to speak, be separated from its external relations. The
motions both of rotation and translation of the system as a
whole can be detached, and those of its various parts amongst
themselves considered.

The question which is put forward for examination is as
follows : If for $\mu = 0$ we have a periodic solution, what are
the conditions necessary in order that the solution shall still
remain periodic when μ is not zero but a small quantity ? It
must be remembered that in this and in what follows, the
term " periodic " has the meaning which has just been given
to it. In order to answer the question, M. Poincaré considers
the system

$$\frac{dx_i}{dt} = X_i$$

where the X_i are functions of the time periodic and of period
2π, as well as of the x_i. Space will not permit me to re-
produce the argument, which finally reduces the answer
to the consideration of the properties of a certain curve
in the neighborhood of the origin. This curve is examined in certain particular cases and notably in the case
where there are an infinite number of periodic solutions for μ
zero, *i.e.* when the period is an arbitrary constant of the
general solution. Generally, it is found that in these cases
the equations *do* admit periodic solutions. In another partic-

ular case, the equations when $\mu = 0$ admit a solution of period 2π, and when μ is small but not zero, save in an exceptional case, the equations admit a solution of period $2k\pi$ (k being a whole number) which is different from the solution of period 2π, and is only not distinct from this latter when μ becomes zero.

If the X_i are periodic with respect to the time, the solution in general if periodic must have the same period. When however the time does not enter into the X_i explicitly, the period of the solution can be anything whatever. Suppose that the period selected when $\mu = 0$ be T. The question resolves itself into finding under what circumstances a solution of period $T + \tau$ is possible when μ is small. The argument proceeds in a somewhat similar manner as in the first case and similar results follow.

To apply these results to the problem of three bodies, suppose $\mu = 0$. Then two of the bodies describe ellipses about the third. At the end of a certain period measured by the difference of their mean motions, the system is found in the same relative position as at the beginning of the period. The solution for $\mu = 0$ is then periodic. Will periodic solutions be still possible when μ, instead of being zero, has a small positive value? From what has been proved above, we can say that such solutions *are* in general possible. M. Poincaré distinguishes three classes :—(1) when the inclinations and eccentricities are zero, (2) when the inclinations only are zero, (3) when the latter are not zero. He then examines these in detail.

Under (1) comes, as a particular case, Dr. Hill's now classic solution, where the mass of one body is supposed to be infinitely great and at an infinite distance, but to have a finite mean motion, and the mass of the other is infinitely small. The solutions are referred to axes moving with the infinitely distant body which takes a circular orbit. The period is one of the arbitraries and can be anything whatever. When $\mu = 0$ the motion is circular, and when μ is small, the curve does not differ much from a circle, and is somewhat elliptical in shape with its shorter axis directed constantly towards the sun. [If the sun be not infinitely distant, the only change in the curve is a loss of symmetry with regard to the line joining the earth and the sun.] Dr. Hill calculated the various shapes which the curve takes for different values of the arbitrary period, corresponding to gradually decreasing values of the constant of vis viva. As this latter constant diminishes, the ratio of the magnitude of the axes becomes greater, until for one particular value of it a cusp appears at each end of the greater axis. This gives what Dr. Hill calls, "the moon of maximum lunation." At the cusp and therefore in quadrature, the moon becomes for a moment stationary with respect to the sun.

He did not pursue the calculations beyond this point. It was stated however that any member of this class of satellites if prolonged beyond the moon of maximum lunation would oscillate to and fro about a mean place in syzygy, never being in quadrature. M. Poincaré points out an inaccuracy in this statement. The satellites which are never in quadrature, are indeed possible but belong to a different class of solution, and are not the analytical continuation of those studied by Dr. Hill. He shows that if we prolonged them beyond the critical orbit, they would cross the line of quadratures six times, cutting their own orbits twice and forming a curve with three closed spaces. The class to which the moons without quadrature belong has, as a limiting case, a moon which is stationary with respect to the sun and which is always either in conjunction or opposition.

M. Poincaré next goes on to consider the canonical system when F_0 is supposed independent of the y_i. This is the general problem of dynamics where the forces depend on the distances only and where we proceed by successive approximations. The first approximation is

$$x_i = const = a_i, \qquad \frac{dy_i}{dt} = const = n_i.$$

If the solution is to be periodic and of period T, all the $n_i T$ must be multiples of 2π. It is then shown that unless the Hessian of F_0 with respect to the x_i vanish, we can have a periodic solution of period T or differing little from T when μ is small. If this Hessian vanish we can sometimes find a function of F_0 whose Hessian does not vanish. If we cannot do this the case must be otherwise examined. Such an examination shows, that when the Hessian of F_0 vanishes, if the mean value R of F_1, with respect to t, admits of a maximum or a minimum, periodic solutions are still possible.

In the problem of three bodies, F_1 corresponds to the disturbing function, and we are led to periodic solutions of the second and third kinds. Here R does admit of a maximum or a minimum, and hence such periodic solutions are always possible. The periodic solutions of the first kind only cease to exist when n' is a multiple of $n - n'$. When, however, this ratio $n' : n - n'$ is nearly a whole number, as happens in several cases in the solar system, a large inequality will exist and its principal part can be calculated suitably by the help of these periodic solutions.

In the next chapter M. Poincaré passes on to the consideration of the characteristic exponents. One solution of a system of differential equations being known, it is required to find a solution differing little from it. The equations of variations are formed in the usual way, and these bring in the equations given above, which involve the characteristic exponents. As an example of the use of these equations, Dr. Hill's work on the motion of the lunar perigee is quoted, where he obtains the principal part of it accurately to a large number of places of decimals.[*]

It is then considered under what circumstances one or more of the exponents become zero, and their effect on the existence of a periodic solution. The argument and result depend chiefly on two things : first, the presence in, or absence from, the X_i of the time explicitly, and, secondly, the existence or non-existence of single-valued integrals of the system. If canonical equations be used, the exponents are equal and opposite in pairs. With the limitation that F_0 does not depend on the y_i, two exponents will be zero, and unless certain conditions be fulfilled, two exponents only will be zero. In the periodic solutions of the problem of three bodies, whether in one plane or not, two exponents and two only are zero. The solutions corresponding to these exponents are called "*solutions dégénérescentes*," and are of the form

$$\xi_i = S_i'', \ \eta_i = T_i'',$$
$$\xi_i = S_i^* + t \, S_i'', \ \eta_i = T_i^* + t \, T_i'',$$

in which the S, T are periodic.

The canonical system given above has an integral which is known, namely the integral of vis viva. The author devotes himself in Chapter V. principally to prove that, save in certain exceptional cases, there does not exist any single-valued algebraic or transcendental integral other than that of vis viva. For this a function Φ is supposed to be analytic and single-valued for all values of x, y, μ within a certain region, and within this region to be developable according to powers of μ, thus :

$$\Phi = \Phi_0 + \mu \, \Phi_1 + \mu^2 \, \Phi_2 + \ . \ .$$

As long as Φ_0 is not a function of F_0, it is proved that $\Phi = const.$ cannot be an integral of the system. If Φ_0 be a function of F_0, it is possible to find another integral which is distinct from F, and which does not reduce to F_0 when μ is zero. In case, however, the Hessian of F_0 be zero, an exceptional case arises, and it is in this exceptional case that the

---

[*] *Acta Mathematica*, Vol. VIII. See also a note by the writer in *Mo. Not. R. A. S.*, Vol. XVII. No. 6.

importance of the principle applied to problems in dynamics is seen. A general set of conditions is found, necessary but not sufficient for the existence of another integral of the equations. These conditions take the form of relations between the co-efficients in the development of F.

Applying these to the problem of three bodies, the author arrives at the conclusion, *that there cannot exist any new transcendental or algebraic single-valued integral of the problem of three bodies other than the well-known ones, whether we consider the particular cases of two, three, or the general case of four degrees of liberty mentioned above.* This important result is of course applicable here to the case only when μ is small, a restriction which nevertheless occurs in most problems of celestial mechanics. It is pointed out, however, that M. Bruns has demonstrated that there cannot exist any other *algebraic* single-valued integral for *any* values of the masses. In actual application M. Poincaré's theorem will be found the more useful, since he includes transcendental as well as algebraic forms in his demonstration.

The most interesting example given to illustrate the general theorem is that of the motion of a solid suspended from a fixed point and acted on by gravity only. The distance of the centre of mass of the body from the point of suspension is supposed small. Two integrals are known : is it possible that a third can exist ?* When the conditions are applied it is found that there is nothing to prevent the existence of a third integral, but since the conditions are necessary and not sufficient nothing proves that it does exist ; such an integral however cannot be algebraic.

Chapters VI. and VII. treat of the disturbing function and M. Poincaré's asymptotic solutions, respectively. In the consideration of the latter a series appears which is divergent in a manner analogous to Sterling's series.

ERNEST W. BROWN.

HAVERFORD COLLEGE, PA., *April*, 1892.

---

* For an elementary discussion of this problem, see Routh's Rigid Dynamics (4th ed.) Vol. II., Chaps. IV., V.

THE RELATIONS OF ANALYSIS AND MATHE–MATICAL PHYSICS.*

AN ADDRESS BEFORE THE INTERNATIONAL CONGRESS OF MATHEMATICIANS, ZÜRICH, AUGUST 9, 1897.

BY PROFESSOR HENRI POINCARÉ.

I.

DOUBTLESS you are often asked what is the utility of mathematics and whether its nicely constructed theories, drawn entirely from the mind, are not artificial products of our caprice.

Among the persons who ask this question, I must make a distinction. The practical class demand of us nothing but means of getting money. These do not deserve to be answered. Rather ought it to be demanded of them what is the good of accumulating so much wealth and whether, in order to have time for its acquisition, it is necessary to neglect art and science, which alone render the soul capable of enjoying it,

Et, propter vitam, vivendi perdere causas.

Moreover, a science produced with a view single to its applications is impossible ; truths are fruitful only if they are concatenated ; if we cleave to those only of which we expect an immediate result, the connecting links will be lacking, and there will be no longer a chain.

The men who are most disdainful of theory find therein, without suspecting it, a daily aliment. Were they deprived of this aliment, progress would quickly be arrested, and we should very soon settle into Chinese immobility.

But we have sufficiently occupied ourselves with the uncompromising practicians; besides these there are those who are curious about Nature only and who ask us if we are in position to help them to a better comprehension of her. In response we have only to show them two monuments, already rough-hewn, celestial mechanics and mathematical physics. They would doubtless concede that these monuments are well worth the labor they have cost. But this is not enough.

Mathematics has a triple end. It should furnish an instrument for the study of nature. Furthermore, it has

*Translated by permission from the *Revue Générale des Sciences*, vol. 8, No. 21, pp. 857–861, by Mr. C. J. KEYSER, Columbia University.

Reprinted from Bull. Amer. Math. Soc. 4 (March 1898), 247–255

a philosophic end, and, I venture to say, an end esthetic. It ought to incite the philosopher to search into the notions of number, space, and time; and, above all, adepts find in mathematics delights analogous to those that painting and music give. They admire the delicate harmony of numbers and of forms; they are amazed when a new discovery discloses for them an unlooked for perspective; and the joy they thus experience, has it not the esthetic character although the senses take no part in it? Only the privileged few are called to enjoy it fully, it is true; but is it not the same with all the noblest arts? Hence I do not hesitate to say that mathematics deserves to be cultivated for its own sake and that the theories not admitting of application to physics deserve to be studied as well as the others.

Even if the physical and esthetic ends were not conjoint, we ought to sacrifice neither the one nor the other. But these two ends are inseparable, and the best means of attaining the one is to aim at the other, or at least never to lose sight of it. a fact I shall now try to demonstrate by showing precisely the nature of the relations between pure science and its applications.

The mathematician ought not to be for the physicist a simple provider of formulæ; between the two there ought to be more intimate collaboration. Mathematical physics and pure analysis are not simply adjacent powers maintaining the relations of good neighborhood; they interpenetrate, and their spirit is the same. This we shall the better comprehend when I shall have shown what physics receives from mathematics and what mathematics, in return, borrows from physics.

II.

The physicist cannot demand of the analyst a revelation of new truth; the analyst can at best aid the physicist in the presentation of truth.

The time is past when people sought to anticipate experience, or to construct the world completely upon certain premature hypotheses. Of all the theories in which they delighted naïvely only a century ago, there remains to-day nothing but ruins.

Now all laws are derived from experience, but, to enunciate them, a special language is needed; ordinary language is too poor; it is besides too vague to express relations so delicate, so complex, and so precise. Here, then, is a prime reason why the physicist cannot dispense with mathematics ; it alone furnishes him with an adequate language.

Neither is it a small matter that a language be fit : not to pass from the domain of physics, the unknown man who invented the word *chaleur* doomed many generations to error ; heat has been treated as a substance simply because it was designated by a substantive, and it was believed to be indestructible. On the other hand, he who invented the word *électricité* has had the unmerited good fortune implicitly to endow physics with a new law, of the conservation of electricity, which, by a pure hazard, is found to be exact, at least up to the present time.

Nay, to pursue the comparison, writers who embellish language, who treat it as an object of art, at the same time make it a suppler instrument, fitter to render the finer shades of thought. We see, therefore, how the analyst, who pursues a purely esthetic end, contributes thus to create a language better suited to the needs of the physicist.

But this is not all ; law issues from experience but it does not do so immediately. Experience is individual, the law derived therefrom is general ; experience is only approximate, law is precise or at least pretends to be. Experience is produced under conditions always complex ; the enunciation of the law eliminates these complications. This is what is termed "elimination of systematic errors."

In a word, in order to derive law from experience, it is necessary to generalize, a necessity that imposes itself upon the most circumspect observer. But how generalize? Every particular truth can evidently be generalized in an infinity of ways. Among the thousand paths that open before us, it is necessary to make a choice, at least a provisional one ; in this choice what shall guide us?

Analogy alone. But how vague this word ! Primitive man takes cognizance of only rude analogies, those that strike the senses, analogies of color and sound. He would never have attempted to connect, for example, light and radiant heat.

What has taught us to discern those genuine, profound analogies that the eye does not see and only the reason divines, if not the mathematical spirit disregarding matter in order to attach itself to pure form? It is this spirit that has directed us to name with the same name things differing in respect to matter only ; to name with the same name, for example, the multiplication of quaternions and that of whole numbers.

If quaternions, of which I just spoke, had not been so

promptly utilized by English physicists, many persons would doubtless have regarded them as only an idle dream; nevertheless, by teaching us to connect things separated by appearances, they would have already rendered us the apter to penetrate the secrets of nature.

Such are the services the physicist may expect from analysis; but in order that this science may render them, it must be cultivated on the largest scale, free from immediate preoccupation with utility; the mathematician must have labored as an artist. What we demand of him is to aid us in seeing, in discerning our way in the labyrinth that presents itself to us. He that sees best is he that has risen to the highest level. Examples abound, I will confine myself to the most striking.

The first will show us how it suffices to change language in order to perceive generalizations that were not at first suspected. When the Newtonian law was substituted for that of Kepler, only the elliptic motion was known. Now, so far as this motion is concerned, the two laws differ in form only; we pass from the one to the other by a simple differentiation. And yet from the law of Newton may be deduced, by an immediate generalization, all the effects of perturbations and the whole of celestial mechanics. Never, on the other hand, had we kept to Kepler's statement, would we have regarded the orbits of the disturbed planets—those complicated curves whose equations have never been written—as the natural generalization of the ellipse. The progress of observation would have served only to produce belief in chaos.

The second example deserves equal consideration. When Maxwell began his labors, the then recognized laws of electrodynamics, accounted for all the known facts. It was not a new experience that invalidated them, but, viewing them under a new aspect, Maxwell saw that the equations became more symmetric on the addition of a term, and, on the other hand, that this term was too small to produce appreciable effects by the old methods. We know that the *a priori* views of Maxwell awaited an experimental confirmation twenty years; or, if you prefer, Maxwell was twenty years in advance of experience. How was this triumph achieved?

Maxwell was profoundly impregnated with the sense of mathematical symmetry; would the case have been the same, had not others, before him, pursued this symmetry for its own beauty?

Maxwell was accustomed "to think in vectors," but vectors are introduced into analysis by the theory of imaginaries.   And those who invented imaginaries little suspected that these would be turned to account in the study of the real world, a fact sufficiently proved by the name they gave them.

Maxwell, in a word, was perhaps not a skilled analyst but this skill would have been for him a useless and cumbrous baggage.   On the other hand, he had in the highest degree a fine sense of mathematical analogies.   On that account he became a thorough mathematical physicist.

The example of Maxwell teaches us yet another thing. How should the equations of mathematical physics be treated?   Ought we simply to deduce from them all the consequences, and regard them as intangible realities?   Far from it : that which they ought especially to teach us is that we can and must transform them, for thus shall we derive something useful from them.

The third example goes to show us how we may perceive mathematical analogies among phenomena which are neither apparently nor really so related physically that the laws of one phenomenon aid us in divining those of the other.   A single equation, that of Laplace, is encountered in the theories of Newtonian attraction, of the motion of liquids, of the electrical potential, of magnetism, of the propagation of heat and in many others besides.   What of it?   These theories seem like images traced the one upon the other. They are mutually illuminated by each appropriating the language of the others ; ask the electricians if they do not felicitate themselves on having invented the phrase, "flux de force," suggested by hydrodynamics and the theory of heat.

Thus mathematical analogies not only enable us to surmise physical analogies but are still useful when these latter are wanting.

To summarize, the end of mathematical physics is not merely to facilitate the numerical calculation of certain constants or the integration of certain differential equations. It is more, it is above all to disclose to the physicist the concealed harmonies of things by furnishing him with a new point of view.

Of all parts of analysis, those are the highest and purest, so to speak, which will be most productive in the hands of such as know how to use them.

III.

Let us now consider what analysis owes to physics.
We should quite ignore the history of science if we failed

to recall how the desire to know nature has constantly exerted upon the development of mathematics their happiest influence.

In the first place, the physicist propounds problems whose solution he awaits at our hands. By proposing them to us he has paid largely in advance for the service we should render by their eventual solution. If I may be permitted to pursue my comparison with the fine arts, the pure mathematician who should forget the existence of the external world would be like a painter who knew how to combine colors and forms harmoniously but who had no models. His creative power would be quickly perverted.

Possible combinations of numbers and symbols form an infinite multitude. How shall we choose from this multitude such as are worthy to detain our attention? Shall we be guided solely by our caprice? This caprice, which, moreover, would ere long wear itself out, would doubtless lead us far asunder, and we should very soon cease to understand each other.

But this is only a minor phase of the question. Physics will doubtless prevent us from going astray, but it will preserve us from a much more formidable danger: that of revolving constantly in the same circle. History proves that physics has not only constrained us to choose among the hosts of problems that present themselves; it has forced upon us those which we had otherwise never attempted. How varied soever the imagination of man, nature is yet a thousand times richer. To pursue her, we should take paths that have been neglected, and these will often conduct to summits whence new landscapes will be revealed.

What more useful! It is with mathematical symbols as with physical realities; by comparing the different aspects of things we shall be able to comprehend their intimate harmony, which alone is beautiful and therefore worthy of our efforts.

The first example I will cite is so old that one is prone to forget it; it is, notwithstanding, the most important of all. The sole natural object of mathematical thought is the whole number. It is the external world that has imposed upon us the continuum, an invention doubtless of our own, but one that the external world forced us to make. Without it, there could be no infinitesimal analysis; all mathematical science would reduce to arithmetic or to the theory of substitutions.

On the other hand, we have devoted to the study of the continuum almost all our time and powers. Who would regret it? Who would pretend that this time and energy have been lost?

Analysis unfolds before us infinite perspectives of which arithmetic has no suspicion; it shows us at a glance a grand assemblage, of simple and symmetric order. In number theory, on the contrary, where the unexpected reigns, the view is, so to speak, arrested at every step.

Doubtless it will be said that, apart from the whole number, there is no rigor and consequently no mathematical truth; that everywhere the whole number is concealed, and that we should try to render the veils transparent, even at the expense of interminable repetition. Let us not be such purists but let us be grateful to the *continuum*, which, if everything proceeds from the whole number, was alone capable of causing so great issue therefrom.

Need I, moreover, recall the fact that Mr. Hermite has derived a striking advantage from the introduction of continuous variables into the theory of numbers? Thus, the domain itself of the whole number has been invaded, and as a result order has been established there where disorder prevailed. Such is our debt to the *continuum* and, by consequence, to physical nature.

The series of Fourier is a precious instrument continually employed by the analyst; but Fourier invented it to solve a physical problem; if his problem had not been naturally set, we should never have dared to render to *discontinuum* its rights; we should for yet a long time have regarded the continuous functions as the only genuine functions.

The notion of *function* has been thereby considerably extended and, at the hands of certain analyst logicians, has received an unforeseen development. These analysts have thus ventured into the regions of the purest abstraction and have departed as far as possible from the real world. It was a physical problem, however, that furnished the point of departure.

After the series of Fourier, other analogous series were introduced into the domain of analysis; they enter by the same gate; they were conceived in view of applications. It is sufficient to cite those that have as elements the sphere functions or the functions of Lamé.

The theory of partial differential equations of the second order has had an analogous history: it was specially developed by and for physics. If the analysts had been left to their natural tendencies, the following is probably how they would have viewed these equations, and how they would have chosen the limiting conditions:

Consider, for example, an equation between two variables x and y and a function F of these two variables. They would have assumed F and $\dfrac{dF}{dx}$ for $x = 0$. This was done, for example, by Mme. de Kowalewski in her celebrated memoir. But there is a host of other ways of putting the problem. F may be given along an entire closed contour, as in the problem of Dirichlet, or the ratio of F to $\dfrac{dF}{dn}$ may be given as in the theory of heat.

It is to physics that we owe all these ways of putting the problem. We may, then, say that without physics the theory of partial differential equations would not be known.

It is useless to multiply examples; I have said enough to warrant a conclusion. When the physicists require of us the solution of a problem, they do not thus impose drudgery upon us; on the contrary, we are under obligation to them.

IV.

But this is not all; physics does not merely furnish problems for solution; it aids us in finding means therefor; and that in two ways: it causes us to surmise the solution, it suggests the proof.

I have mentioned above the fact that the equation of Laplace is met with in a host of far separated physical theories. We find it again in geometry in the theory of conformal representation, and in pure analysis in that of imaginaries. Thus in the study of functions of complex variables the analyst, besides the geometric image which is his usual instrument, finds several physical images that can be used with equal success. Thanks to these images he can see at a glance what pure deduction could show him only in succession. He thus collects the separate elements of the solution, and by a sort of intuition divines before he can demonstrate.

Divine before demonstrate! Need I recall the fact that all important discoveries are thus made? What truths of which the physical analogies give us a presentiment, and which we are not yet in position to establish by rigorous argument! For example, mathematical physics introduces a great number of developments in series. That these series converge, no one doubts; but mathematical certainty is wanting. These are so many assured conquests for the investigators who shall come after us.

Physics, on the other hand, does not merely furnish us with solutions; it also in certain measure provides arguments It will be sufficient to recall how Mr. Klein, in a question respecting Riemann surfaces, had recourse to the properties of electric currents.   Arguments of this class, it is true, are not rigorous in the sense that the analyst attaches to this word.

And, in this connection, arises the question : How a demonstration that is not rigorous enough for the analyst, can be sufficient for the physicist?   It seems there can not be two rigors, that rigor is or it is not, and that there where it is not, argument can not exist.

We will best comprehend this apparent paradox by recalling the conditions of the applicability of number to natural phenomena.   Whence arise, in general, the difficulties encountered when one seeks to give a rigorous demonstration? One strikes them almost always when attempting to establish that such a quantity tends towards such a limit, or that such a function is continuous, or that it has a derivative. Now, the numbers that the physicist measures by experience are known to him only approximately, and, on the other hand, any function whatever always differs by as little as we please from a discontinuous function, and, at the same time, it differs as little as we please from a continuous function.

The physicist may accordingly suppose, at pleasure, that the function studied is continuous or that it is discontinuous ; that it has a derivative or that it has not ; and this, without fear of contradiction by either present or future experience.   One understands how, with such liberty, he makes play of the difficulties that detain the analyst.   He may always reason as if all the functions occurring in his calculations were entire polynomials.

Thus the view that suffices for physics is not such reasoning as analysis requires.   It does not follow that the one is not able to aid the other.

So many physical observations have been already transformed into rigorous demonstrations that this transformation is easy to-day.   Examples abound, if I did not fear, in citing them, to weary your attention, and if this conference had not been already too long.

I hope I have said enough to show that pure analysis and mathematical physics may be reciprocally helpful without either entailing sacrifice upon the other, and that each should rejoice in whatever exalts its associate.

BULLETIN (New Series) OF THE
AMERICAN MATHEMATICAL SOCIETY
Volume 37, Number 1, Pages 25–38
S 0273-0979(99)00801-0
Article electronically published on December 21, 1999

THE PRESENT AND THE FUTURE
OF MATHEMATICAL PHYSICS

PROFESSOR H. POINCARÉ

What is the present state of mathematical physics? What are its problems? What is its future? Is it about to change its orientation? Will the object and methods of this science appear in ten years to our immediate successors in the same light as they appear to us? Or are we to witness a far-reaching transformation? These are the questions we are forced to face to-day at the outset of our inquiry.

It is easy to ask; difficult to answer. If we felt tempted to hazard a prediction, we should easily resist this temptation by stopping to think of the nonsense the most eminent scholars of a hundred years ago would have spoken in answer to the question of what this science would be in the nineteenth century. They would have thought themselves bold in their predictions; and after the event how timid we should have found them! Do not expect of me therefore any kind of prophesy.

But if, like all prudent physicians, I refuse to give a prognosis, still I cannot deny myself a little diagnosis. Well, then, yes; there are symptoms of a serious crisis, which would seem to indicate that we may expect presently a transformation. However, there is no cause for great anxiety. We are assured that the patient will not die, and indeed we may hope that this crisis will be salutary, since the history of the past would seem to insure that. In fact, this crisis is not the first, and in order to understand it it is well to recall those which have gone before. Allow me a brief historical sketch.

Mathematical physics, as we are well aware, is an offspring of celestial mechanics, which gave it birth at the end of the eighteenth century at the moment when it had itself attained its complete development. The child, especially during its first years, showed a striking resemblance to its mother.

The astronomical universe consists of masses, undoubtedly of great magnitude, but separated by such immense distances that they appear to us as material points; these points attract each other in the inverse ratio of the squares of their distances, and this attraction is the only force which affects their motion. But if our senses were keen enough to show us all the details of the bodies which the physicist studies, the spectacle thus disclosed would hardly differ from the one which the astronomer contemplates. There too we should see material points separated by intervals which are enormous in comparison with their dimensions, and describing orbits according to regular laws. These infinitesimal stars are the atoms. Like the stars proper, they attract each other or repel, and this attraction or repulsion, which is along

2000 *Mathematics Subject Classification.* Primary 83A05, 70A05.

Reprinted from Bull. Amer. Math. Soc. **12** (1906), 240–260.

Translated with the author's permission by Professor J. W. Young.

Address delivered before the Section of Applied Mathematics of the International Congress of Arts and Science, St. Louis, September 24, 1904.

the line joining them, depends only on the distance. The law according to which this force varies with the distance is perhaps not the law of Newton, but it is analogous thereto: instead of the exponent -2 we probably have another exponent, and from this diversity in the exponents proceeds all the diversity of the physical phenomena, the variety in qualities and sensations, all the world of color and sound which surrounds us; in a word, all nature.

Such is the primitive conception in its utmost purity. Nothing remains but to inquire in the different cases, what value must be given to this exponent in order to account for all the facts. On this model, for example, Laplace constructed his beautiful theory of capillarity; he simply regards the latter as a special case of attraction, or, as he says, of universal gravitation, and no one is surprised to find it in the middle of one of the five volumes of his celestial mechanics. More recently Briot believes he has laid bare the last secret of optics, when he has proved that the atoms of the ether attract each other in the inverse sixth power of the distance; and does not Maxwell, Maxwell himself, say somewhere that the atoms of a gas repel each other in the inverse ratio of the fifth power of the distance? We have the exponent -6 or -5, instead of the exponent -2; but it is always an exponent.

Among the theories of this period there is a single one that forms an exception, namely that of Fourier; here there are indeed atoms acting at a distance; they send each other heat, but they do not attract each other, they do not stir. From this point of view, Fourier's theory must have appeared imperfect and provisional to the eyes of his contemporaries, and even to his own.

This conception was not without greatness; it was alluring, and many of us have not definitely given it up; they know that the ultimate elements of things will not be attained, except by disentangling with patience the complex skein furnished us by our senses; that progress should be made step by step without neglecting any intermediate portions; that our fathers were unwise in not wishing to stop at all the stations; but they believe that when we once arrive at these ultimate elements, we shall meet again the majestic simplicity of celestial mechanics.

Nor has this conception been useless; it has rendered us a priceless service inasmuch as it has contributed to making more precise the fundamental concept of the physical law. Let me explain: What did the ancients understand by a law? It was to them an internal harmony, statical as it were, and unchangeable; or else a model which nature tried to imitate. To us a law is no longer that at all; it is a constant relation between the phenomenon of to-day and that of to-morrow; in a word, it is a differential equation.

Here we have the ideal form of the physical law; and, indeed, it is Newton's law which first gave it this form. If, later on, this form has become inured in physics, it has become so precisely by copying as far as possible this law of Newton, by using celestial mechanics as a model.

Nevertheless there came a day when the conception of central forces appeared no longer to suffice, and this is the first of the crises to which I referred a moment ago.

What was done? Abandoned was the thought of exploring the details of the universe, of isolating the parts of this vast mechanism, of analyzing one by one the forces which set them going; and one was content to take as guides certain general principles which have precisely the object of relieving us of this minute study. How is this possible? Suppose we have before us any kind of machine; the part of the mechanism where the power is applied and the ultimate resultant motion alone are visible, while the transmissions, the intermediate gearing whereby the motion is communicated from one part to another, are hidden in the interior and escape our notice; we know not whether the transmissions are made by cog-wheels or by belts, by connecting-rods or other contrivances. Shall we say that it is impossible for us to learn anything about this machine unless we are allowed to take it apart? You well know that such is not the case, and that the principle of the conservation of energy suffices to furnish us the most interesting feature. We can easily show that the last wheel turns ten times more slowly than the first, since these two wheels are visible; and we can conclude therefrom that a couple applied to the first will be in equilibrium with a couple ten times as great applied to the second. To obtain this result, it is in no wise necessary to look into the mechanism of this equilibrium, or to know how the forces balance in the interior of the machine; it is sufficient to make sure that it is impossible for this balancing not to take place.

Very well! In the case of the universe, the principle of the conservation of energy can render us the same service. This universe also is a machine, much more complicated than any in use in the industries, of which nearly all the parts are deeply hidden; but by observing the motion of those which we can see, we can by the aid of this principle draw conclusions which will remain valid no matter what the details of the invisible mechanism which actuates them.

The principle of the conservation of energy, or Mayer's principle, is certainly the most important, but it is not the only one; there are others from which we can derive the same advantage. These are:

Carnot's principle, or the principle of the dissipation of energy.

Newton's principle, or the principle of the equality of action and reaction.

The principle of relativity, according to which the laws of physical phenomena must be the same for a stationary observer as for one carried along in a uniform motion of translation, so that we have no means, and can have none, of determining whether or not we are being carried along in such a motion.

The principle of the conservation of mass, or Lavoisier's principle. I will add the principle of least action.

The application of these five or six general principles to the various physical phenomena suffices to teach us what we may reasonably hope to know about them. The most remarkable example of this new mathematical physics is without doubt Maxwell's electro-magnetic theory of light. What is the ether? How are its molecules distributed? Do they attract or repel each other? Of these things we know nothing. But we know that this medium transmits both optical and electrical disturbances; we know that this transmission must take place in conformity with the general principles of mechanics and that suffices to establish the equations of the electro-magnetic field.

These principles are the boldly generalized results of experiment; but they appear to derive from their very generality a high degree of certainty. In fact, the greater the generality, the more frequent are the opportunities for verifying them, and such verifications, as they multiply, as they take the most varied and most unexpected forms, leave in the end no room for doubt.

Such is the second phase of the history of mathematical physics, and we have not yet left it. Shall we say that the first has been useless, that for fifty years science was on a wrong path and that there is nothing to do but to forget all that accumulation of effort which a vicious conception from the very beginning doomed to failure? By no means! Do you think the second period could have existed without the first?

The hypothesis of central forces contained all the principles; it involved them as necessary consequences; it involved the principle of the conservation of energy, as well as that of mass, and the equality of action and reaction, and the law of least action, which appeared to be sure, not as experimental facts, but as theorems, and of which the statement had I know not how much greater precision and lesser generality than under their present form.

It is the mathematical physics of our fathers which has gradually made us familiar with these various principles, which has taught us to recognize them in the different garbs in which they are disguised. They have been compared with the results of experiment; it has been found necessary to change their expression in order to make them conform to the facts; thus they have been extended and strengthened. In this way they came to be regarded as experimental truths. The conception of central forces then became a useless support, or rather an encumbrance, inasmuch as it imposed upon the principles its own hypothetical character.

The bounds then are not broken, because they were elastic; but they have been extended. Our fathers who established them have not labored in vain; and in the science of to-day we recognize the general features of the outline they traced.

Are we now about to enter upon a third period? Are we on the eve of a second crisis? Are these principles on which we have reared everything about to fall in their turn? This has recently become a vital question.

Hearing me speak thus, you are thinking without doubt of radium, that great revolutionary of the present day; and indeed I shall return to it presently. But there is something else. It is not merely the conservation of energy that is concerned; all the other principles are in equal danger, as we shall see by successively passing them in review.

Let us begin with Carnot's principle. It is the only one which does not present itself as an immediate consequence of the hypothesis of central forces; quite to the contrary, indeed, it appears, if not actually to contradict this hypothesis, at least not to be reconcilable with it without some effort. If physical phenomena were due exclusively to the motion of atoms the mutual attractions of which depend only on the distance, it would seem that all these phenomena should be reversible; if all the initial velocities were reversed, these atoms, if still subject to the same forces, should traverse their trajectories in the opposite direction, just as the earth would describe backward this same elliptical orbit that it now describes forward, if the initial conditions of its motion had been reversed. Thus, if a physical phenomenon is possible, the inverse phenomenon should be equally possible, and one should be able to retrace the course of time. Now, it is not so in nature, and this it is precisely that the principle of Carnot teaches us; heat may pass from a hot body to a cold; it is impossible to compel it to take the opposite route and to re-establish differences of temperature which have disappeared. Motion can be entirely destroyed and transformed into heat by friction; the converse transformation can only occur partially.

Efforts have been made to reconcile this apparent contradiction. If the world tends toward uniformity, it is not because its ultimate parts, though diversified at the start, tend to become less and less different; it is because moving at random they become mixed. To an eye which could distinguish all the elements, the variety would remain always as great; every grain of this powder retains its originality and does not fashion itself after its neighbors; but as the mixture becomes more and more perfect, our rough senses perceive only uniformity. That is why, for example, temperatures tend to equalize themselves, without its being possible to go back.

A drop of wine, let us say, falls into a glass of water; whatever the internal motion of the liquid, we shall soon see it assume a uniformly roseate hue, and from then on no possible shaking of the vessel would seem to be capable of again separating the wine and the water. Here, then, we have what may be the type of the irreversible phenomenon of physics: to hide a grain of barley in a great mass of wheat would be easy; to find it again and to remove it is practically impossible. All this has been explained by Maxwell and Boltzmann, but the man who has put it most clearly was Gibbs, in a book too little read because it is a little difficult to read, in his Elements of Statistical Mechanics.

To those who take this point of view, Carnot's principle is an imperfect principle, a sort of concession to the frailty of our senses; it is because our eyes are too coarse that we do not distinguish the elements of the mixture; it is because our hands are too coarse that we cannot compel them to separate; the imaginary demon of Maxwell, who can pick out the molecules one by one, would be quite able to constrain the world to move backwards. That it should return of its own accord is not impossible; it is only infinitely improbable; the chances are that we should wait a long time for that combination of circumstances which would permit a retrogression; but, sooner or later, they will occur, after years, the number of which would require millions of figures. These reservations, however, all remained theoretical; they caused little uneasiness and Carnot's principle preserved all of its practical value.

But now here is where the scene changes. The biologist, armed with his microscope, has for a long time noticed in his preparations certain irregular motions of small particles in suspension; this is known as Brown's motion. He believed at first that it was a phenomenon of life, but he soon saw that inanimate bodies hopped about with no less ardor than others; he then turned the matter over to the physicists. Unfortunately, the physicists did not become interested in the question for a long time. Light is concentrated, so they argued, in order to illuminate the microscopical preparation; light involves heat, and this causes differences in temperature and these produce internal currents in the liquid, which bring about the motions referred to.

M. Gouy had the idea of looking a little more closely, and thought he saw that this explanation was untenable; that the motion becomes more active as the particles become smaller, but that they are uninfluenced by the manner of lighting. If, then, these motions do not cease, or, rather, if they come into existence incessantly, without borrowing from any external source of energy, what must we think? We must surely not abandon on this account the conservation of energy; but we see before our eyes motion transformed into heat by friction and conversely heat changing into motion, and all without any sort of loss, since the motion continues forever. It is the contradiction of Carnot's principle. If such is the case, we need no longer the infinitely keen eye of Maxwell's demon in order to see the world move backward; our microscope suffices. The larger bodies, those of a tenth of a millimeter, for example, are bombarded from all sides by the moving atoms, but they do not stir, because these shocks are so numerous that the law of probabilities requires them to compensate each other; but the smaller particles are hit too rarely to have this compensation take place with any degree of certainty and are thus incessantly tossed about. And so one of our principles is already in danger.

Let us consider the principle of relativity; this principle is not only confirmed by our daily experience, not only is it the necessary consequence of the hypothesis of central forces, but it appeals to our common sense with irresistible force. And yet it also is being fiercely attacked. Let us think of two electrified bodies; although they seem to be at rest, they are, both of them, carried along with the motion of the earth; Rowland has shown us that an electric charge in motion is equivalent to a current; these two charged bodies, then, are equivalent to two parallel currents in the same direction; these two currents should attract each other. By measuring this attraction we should be measuring the velocity of the earth; not its velocity relative to the sun and stars, but its absolute velocity.

I know what will be said; it is not its absolute velocity; it is its velocity relative to the ether. But, how unsatisfactory that is! Is it not clear that with this interpretation, nothing could be inferred from the principle? It could no longer teach us anything, simply because it would no longer fear any contradiction. Whenever we have succeeded in measuring anything, we would always be free to say that it is not the absolute velocity, and if it is not the velocity relative to the ether, it might always be the velocity relative to some new unknown fluid with which we might fill all space.

And then experiment, too, has taken upon itself to refute this interpretation of the principle of relativity; all the attempts to measure the velocity of the earth relative to the ether have led to negative results. Herein experimental physics has been more faithful to the principle than mathematical physics; the theorists would have dispensed with it readily in order to harmonize the other general points of view; but experimentation has insisted on confirming it. Methods were diversified; finally Michelson carried precision to its utmost limits; nothing came of it. It is precisely to overcome this stubborness that to-day mathematicians are forced to employ all their ingenuity.

Their task was not easy, and if Lorentz has succeeded, it is only by an accumulation of hypotheses.

The most ingenious idea is that of local time. Let us imagine two observers, who wish to regulate their watches by means of optical signals; they exchange signals, but as they know that the transmission of light is not instantaneous, they are careful to cross them. When station B sees the signal from station A, its timepiece should not mark the same hour as that of station A at the moment the signal was sent, but this hour increased by a constant representing the time of transmission. Let us suppose, for example, that station A sends it signal at the moment when its time-piece marks the hour zero, and that station B receives it when its time-piece marks the hour t. The watches will be set, if the time t is the time of transmission, and in order to verify it, station B in turn sends a signal at the instant when its time-piece is at zero; station A must then see it when its time-piece is at t. Then the watches are regulated.

And, indeed, they mark the same hour at the same physical instant, but under one condition, namely, that the two stations are stationary. Otherwise, the time of transmission will not be the same in the two directions, since the station A, for example, goes to meet the disturbance emanating from B, whereas station B flees before the disturbance emanating from A. Watches regulated in this way, therefore, will not mark the true time; they will mark what might be called the local time, so that one will gain on the other. It matters little, since we have no means of perceiving it. All the phenomena which take place at A, for example, will be behind time, but all just the same amount, and the observer will not notice it since his watch is also behind time; thus, in accordance with the principle of relativity he will have no means of ascertaining whether he is at rest or in absolute motion. Unfortunately this is not sufficient; additional hypotheses are necessary. We must admit that the moving bodies undergo a uniform contraction in the direction of the motion. One of the diameters of the earth, for example, is shortened by $1/200000000$ as a result of our planet's motion, whereas the other diameter preserves its normal length. Thus we find the last minute differences accounted for. Then there is still the hypothesis concerning the forces. Forces, whatever their origin, weight as well as elasticity, will be reduced in a certain ratio in a world endowed with a uniform translatory motion; or rather that would happen for the components at right angles to the direction of translation; the parallel components will not change. Let us then return to our example of the two electrified bodies; they repel each other; but at the same time, if everything is carried along in a uniform transition, they are equivalent to two parallel currents in the same direction, which attract each other.

This electrodynamic attraction is, then, subtracted from the electrostatic repulsion, and the resultant repulsion is weaker than if the two bodies had been at rest. But since we must, in order to measure this repulsion, balance it by another force, and since all these other forces are reduced in the same ratio, we observe nothing. Everything, then, appears to be in order. But have all doubts been dissipated? What would happen if we could communicate by signals other than those of light, the velocity of propagation of which differed from that of light? If, after having regulated our watches by the optimal method, we wished to verify the result by means of these new signals, we should observe discrepances due to the common translatory motion of the two stations. And are such signals inconceivable, if we take the view of Laplace, that universal gravitation is transmitted with a velocity a million times as great as that of light?

Thus the principle of relativity has in recent times been valiantly defended; but the very vigor of the defense shows how serious was the attack.

And now let us speak of the principle of Newton, concerning the equality of action and reaction. This principle is intimately connected with the preceding and it would seem that the fall of one would involve the fall of the other. Nor must we be surprised to find here again the same difficulties.

The electrical phenomena, it is thought, are due to displacements of small charged particles called electrons which are immersed in the medium we call ether. The motions of these electrons produce disturbances in the surrounding ether; these disturbances are propagated in all directions with the velocity of light, and other electrons initially at rest are displaced when the disturbance reaches the portions of the ether in which they lie. The electrons, then, act one upon the other, but this action is not direct; it takes place by mediation of the ether. Under these conditions, is it possible to have equality between action and reaction, at least for an observer who takes account only of the motion of matter, that is of the electrons, and who ignores that of the ether which he is unable to see? Evidently not. Even if the compensation were exact, it could not be instantaneous. The disturbance is propagated with a finite velocity; it reaches the second electron, therefore, only after the first has long been reduced to rest. This second electron will, then, after an interval, be subjected to the action of the first, but will certainly not at that moment react upon it, since there is no longer anything in the neighborhood of this first electron that stirs.

The analysis of the facts will allow us to become more definite. Let us imagine, for example, a Hertzian oscillator such as those used in wireless telegraphy; it sends energy in all directions; but we may attach to it a parabolic mirror, as was done by Hertz with his smallest oscillators, so as to send all the energy produced in a single direction. What then will happen according to the theory? Why, the apparatus will recoil as though it were a cannon and the projected energy a ball, and that contradicts the principle of Newton, since our present projectile has no mass; it is not matter, it is energy. It is the same, moreover, in the case of a light-house having a reflector, since light is merely a disturbance in the electro-magnetic field. This light-house would recoil, as though the light it sends forth were a projectile. What is the force that must produce this recoil? It is what is known as the Maxwell-Bartholdi pressure; it is very small, and to put it in evidence caused much trouble, even with the most sensitive radiometers; but it is sufficient for our purpose that it exists.

If all the energy issuing from our oscillator strikes a receiver, the latter will act as though it had received a physical shock, which in a sense will represent the compensation of the oscillator's recoil; the reaction will be equal to the action, but they will not be simultaneous; the receiver will advance, but not at the instant when the oscillator recoils. If the energy is propagated indefinitely without meeting a receiver, the compensation will never take place.

Shall we say that the space which separates the oscillator from the receiver and which the disturbance must traverse in passing from one to the other, is not empty, but is filled not only with ether, but with air, or even in inter-planetary space with some subtile, yet ponderable fluid; that this matter receives the shock, as does the receiver, at the moment the energy reaches it, and recoils, when the disturbance leaves it? That would save Newton's principle, but it is not true. If the energy during its propagation remained always attached to some material substratum, this matter would carry the light along with it and Fizeau has shown, at least for the air, that there is nothing of the kind. Michelson and Morley have since confirmed this. We might also suppose that the motions of matter proper were exactly compensated by those of the ether; but that would lead us to the same considerations as those made a moment ago. The principle, if thus interpreted, could explain anything, since whatever the visible motions we could imagine hypothetical motions to compensate them. But if it can explain anything, it will allow us to foretell nothing; it will not allow us to choose between the various possible hypotheses, since it explains everything in advance. It therefore becomes useless.

And then the suppositions that must be made concerning the motions of the ether are not very satisfactory. If the electric charges were doubled, it would be natural to suppose that the velocities of the atoms of the ether also became twice as great, and for the compensation it would be necessary that the mean velocity of the ether become four times as great.

This is why I have for a long time thought that these consequences of the theory, which contradict Newton's principle, would some day be abandoned; and yet the recent experiments on the motion of the electrons emitted by radium seem rather to confirm them.

I now come to Lavoisier's principle concerning the conservation of mass. This is certainly a principle which cannot be tampered with without shaking the science of mechanics. And still there are persons who think that it seems true to us only because in mechanics we consider only moderate velocities, and that it would cease to be so for bodies having velocities comparable with that of light. Now, such velocities are at present believed to have been realized; the cathode rays and those of radium would seem to be formed of very minute particles or electrons that move with velocities that are no doubt less than that of light, but which appear to be about one tenth or one third of it.

These rays can be deflected either by an electric or by a magnetic field, and by comparing these deflections it is possible to measure both the velocity of the electrons and their mass (or rather the ratio of their mass to their charge). But it was found that as soon as these velocities approached that of light a correction was necessary. Since these particles are electrified, they cannot be displaced without disturbing the ether; to put them in motion, it is necessary to overcome a double inertia, that of the particle itself and that of the ether. The total or apparent mass that is measured is then composed of two parts: the real or mechanical mass of the particle and the electrodynamic mass representing the inertia of the ether.

Now, the calculations of Abraham and the experiments of Kaufmann have shown that the mechanical mass properly so called is nothing, and that the mass of the electrons, at least of the negative electrons, is purely of electrodynamic origin. This is what compels us to change our definition of mass; we can no longer distinguish between the mechanical mass and the electrodynamic mass, because then the first would have to vanish; there is no other mass than the electrodynamic inertia; but in this case, the mass can no longer be constant; it increases with the velocity; and indeed it depends on the direction, and a body having a considerable velocity will not oppose the same inertia to forces tending to turn it off its path that it opposes to those tending to accelerate or retard its motion.

There is indeed another resource: the ultimate elements of bodies are electrons, some with a negative charge, others with a positive charge. It is understood that the negative electrons have no mass; but the positive electrons, from what little is known of them, would seem to be much larger. They perhaps have besides their electrodynamic mass a true mechanical mass. The real mass of a body would then be the sum of the mechanical masses of its positive electrons, the negative electrons would not count; the mass defined in this way might still be constant.

Alas, this resource is also denied. Let us recall what we said concerning the principle of relativity and the efforts made to save it. And it is not only a principle that is to be saved; the indubitable results of Michelson's experiments are involved. And so, as was above seen, Lorentz, to account for these results, was obliged to suppose that all forces, whatever their origin, are reduced in the same ratio in a medium having a uniform translatory motion. But that is not sufficient; it is not enough that this should take place for the real forces, it must also be the same in the case of the forces of inertia; it is necessary, therefore—so he says—that *the masses of all particles be influenced by a translation in the same degree as the electromagnetic masses of the electrons.*

Hence, the mechanical masses must vary according to the same laws as the electrodynamic; they can then not be constant.

Do I need to remark that the fall of Lavoisier's principle carries with it that of Newton's? The latter implies that the center of gravity of an isolated system moves in a straight line; but if there no longer exists a constant mass, there no longer exists a center of gravity; indeed the phrase would be meaningless. This is why I said above that the experiments on cathode rays seemed to justify the doubts of Lorentz concerning Newton's principle.

From all these results, if they were to be confirmed, would issue a wholly new mechanics which would be characterized above all by this fact, that there could be no velocity greater than that of light,[1] any more than a temperature below that of absolute zero. For an observer, participating himself in a motion of translation of which he has no suspicion, no apparent velocity could surpass that of light, and this would be a contradiction, unless one recalls the fact that this observer does not use the same sort of timepiece as that used by a stationary observer, but rather a watch giving the "local time."

Here we are then face to face with a question, of which I shall confine myself to the mere statement. If there is no longer any mass what becomes of Newton's law?

Mass has two aspects: it is at the same time a coefficient of inertia and an attracting mass entering as a factor into Newton's law of attraction. If the coefficient of inertia is not constant, can the attracting mass be constant? This is the question.

The principle of the conservation of energy at least still remained and appeared more finely established. Shall I recall to your minds how it too was thrown into discredit? That event made more noise than the preceding; the journals are full of it. Ever since the first work of Becquerel, and above all after the Curies had discovered radium, it was seen that every radioactive substance was an inexhaustible source of radiation. Its activity seemed to continue without change through months and years. That is already a strain on the principles; these radiations in fact were energy, and from the same piece of radium came forth this energy and it came forth indefinitely. But these quantities of energy were too minute to be measured; at least that was the belief, and the matter caused little uneasiness.

The scene changed when Curie thought of placing the radium in a calorimeter. It was then seen that the quantity of heat continuously generated was very considerable.

The explanations advanced were numerous; but in a case of this kind it is not possible to say that an abundance of good does no harm: as long as one explanation has not displaced the others we can not be sure that any one of them is good. For some time, however, one of these explanations seems to be gaining the upper hand and we may reasonably hope that we hold the key to the mystery.

Sir W. Ramsey has attempted to show that radium is transformed, that it contains an enormous amount of energy, but not an inexhaustible amount. The transformation of radium must then produce a million times as much heat as any known transformation; the radium would be exhausted in 1250 years; that is not long, but you see that we are at least sure of being bound to the present state of affairs for some hundreds of years. While we wait our doubts subsist.

In the midst of such ruin, what remains standing? The principle of least action up to now is intact, and Larmor appears to think that it will long survive the others. It is in fact more vague and even more general.

In the presence of this general collapse of principles, what attitude should mathematical physics take? First of all, before becoming too excited, it is well to ask whether all this is really true. All this disparagement of principles is encountered only in the case of the infinitely small; the microscope is needed to see Brown's motion, the electrons are rather tiny, radium is very rare and never more than a few milligrams are together; and then we can ask whether by the side of the minute

---

[1] Because bodies would oppose an increasing inertia to the causes that would tend to accelerate their motion; and when approaching the velocity of light, this inertia would become infinite.

thing that was observed, there was not another minute thing which was not noticed and which counterbalanced the first.

The question is surely debatable, and apparently only experiment can solve it. We should merely have to turn the matter over to the experimenters and, while waiting for them definitely to settle the controversy, not to trouble ourselves with these disquieting problems, and to keep quietly at our work, as though the principles were still unchallenged. We certainly have enough to do without leaving the domain where they can be applied with all certainty; we have enough to keep us busy during this period of doubt.

And yet is it really true that we can do nothing to relieve science of these doubts? It must indeed be said that it has not been experimental physics alone that has brought them into existence; mathematical physics has contributed its share. It was the experimenters who saw radium emit energy; but the theorists were the ones who brought to light all the difficulties inherent in the propagation of light through a moving medium; had it not been for them, they probably would not have been noticed. They have, then, done their best to embarrass us; it is no more than just that they should help us to extricate ourselves.

They must subject to a searching criticism all the new conceptions that I have outlined to-day, nor must they abandon the principles except after a loyal effort to save them. What can they do in this direction? That is what I shall seek to explain.

Among the most interesting problems of mathematical physics a place should be set apart for those that belong to the kinetic theory of gases.

Much has already been done toward their solution, but much remains to do. This theory is an everlasting paradox. We have reversibility in the premises and irreversibility in the conclusions, and a deep chasm between the two. Will statistical considerations, and the law of large numbers, suffice to fill it up? Many points still remain obscure to which it will be necessary to return, and that without doubt several times. In clearing them up, the meaning of Carnot's principle will be better understood, and its general position in dynamics; and we shall be better able to interpret the curious experiment of Gouy to which I referred above.

Should we not also make an effort to obtain a more satisfactory theory of the electrodynamics of moving bodies? It is here, above all, as I indicated sufficiently a short time ago, that the difficulties accumulate; even though we heap up hypotheses, we can not satisfy all the principles at once; no one has succeeded so far in saving some without sacrificing others. But all hope of obtaining better results is not yet lost. Let us, then, take the theory of Lorentz. Let us turn it over and over, let us modify it little by little, and all will be well, perhaps.

Indeed, instead of supposing that bodies in motion undergo a contraction in the direction of motion and that this contraction is the same whatever the nature of these bodies and the forces to which they are subjected, could not a simpler and more natural hypothesis be made? One might suppose, for example, that it is the ether which changes when it is in relative motion with respect to the material substance which passes through it; that, when thus modified, it no longer transmits the disturbances with the same velocity in all directions. It would transmit more rapidly those disturbances which are being propagated parallel to the motion of the substance, be it in the same direction or in the opposite, and less rapidly those which are propagated at right angles. The wave surfaces would then no longer be spheres, but ellipsoids, and one could do without this extraordinary contraction of all bodies.

I am giving this only by way of example, since the modifications which could be tried are evidently susceptible of infinite variation.

It is possible also that astronomy may some day furnish us with data on this point: she it was, in fact, who raised the question by making known to us the phenomenon of the aberration of light. If the theory of the aberration of light is roughly constructed, a curious result is arrived at. The apparent positions of the stars differ from their real positions by reason of the earth's motion, and since this motion is variable the apparent positions vary. The real position we are unable to ascertain, but we can observe the variations of the apparent position. Observations on aberration then show us, not the earth's motion, but the variations of this motion. They can, therefore, teach us nothing concerning the absolute motion of the earth.

These at any rate are the facts under a first approximation; but such would no longer be the case if we could observe the thousandth part of a second. It would then be seen that the variation in the apparent motion of the star depends not only on the variation in the earth's motion, a variation which is well known, since it is the motion of our globe in its elliptical orbit, but also on the mean value of this motion, so that the constant of aberration would not be quite the same for all stars, and that the differences would make known to us the earth's absolute motion in space.

This would be, in another form, the end of the principle of relativity. We are far, it is true, from being able to observe the thousandth part of a second, but after all, say some, the earth's total absolute velocity is perhaps much greater than its velocity relative to the sun; if it were for instance 300 km. per second, instead of 30 km., that would suffice to make the phenomenon observable.

I believe that by reasoning in this manner we carry simplicity in the theory of aberration too far. Michelson has shown, as I have said, that the methods of physics are powerless to put absolute motion in evidence; I am convinced that in the case of astronomical methods it will be the same, no matter how far precision may be carried.

However that may be, the data which astronomy will furnish in this direction will one day be valuable to the physicist. In the meantime I believe that the theorists, keeping in mind the experiments of Michelson, may count on a negative result, and that they would do useful work by constructing a theory of aberration which takes account of it in advance.

But let us return to the earth. There, too, we can help the experimenters. We can, for example, prepare the way by studying thoroughly the dynamics of the electrons; not, be it well understood, by staring from a single hypothesis, but by multiplying the hypotheses as much as possible. It would then be the part of the physicist to use our work in searching for the crucial experiment which would decide between them.

This dynamics of the electrons can be approached from many sides; but among the roads that lead there, there is one which has been somewhat neglected, and yet it is one of those that promise us the most surprises. It is the motion of electrons that produces the lines of the spectrum; this is proved by the phenomenon of Zeemann; what vibrates in an incandescent body is affected by a magnet, and is hence electrified. This is a first very important point; but no one has gone into the question any further. Why are the lines of the spectrum distributed according to a regular law? These laws have been studied by the experimenters in the greatest detail; they are very precise and comparatively simple. A first study of these arrangements brings to mind the harmonics encountered in acoustics; but the difference is great. Not only are the numbers of vibrations not successive multiples of a single number, but we even find nothing analogous to the roots of those transcendental equations, to which we are led by so many problems of mathematical physics: that of the vibrations of an elastic body of any shape, of the Hertzian oscillations in a generator of any form, the problem of Fourier on the cooling of a solid body.

The laws are simpler, but they are of an entirely different kind. To mention only one of these differences, for harmonics of high order the number of vibrations tends toward a finite limit, instead of increasing indefinitely.

This has not yet been explained, and I believe that here is one of the most important of nature's secrets. Lindemann has made a praiseworthy attempt, but in my opinion without success. This attempt should be renewed. We shall thus penetrate, so to speak, into the intimacies of matter. And, from the particular point of view that we occupy to-day, when we shall know why the vibrations of incandescent bodies differ in this way from the vibrations of ordinary elastic bodies, why the electrons do not behave like the matter with which we are familiar, we shall better understand the dynamics of the electrons and it will perhaps be easier for us to reconcile them with the principles.

Now suppose that all these efforts should fail (and when all is said, I do not believe they will), what should be done? Should we seek to rebuild these shattered principles by one stroke, as it were? That, evidently, is always possible, and I take back nothing of what I once said. "Did you not write," you might say, if you were seeking a quarrel with me, "did you not write that the principles, though they are of experimental origin, are now beyond the possibility of experimental attack, because they have become conventions? And now you come to tell us that the triumphs of the most recent experiments put these principles in danger."

Very well, I was right formerly, and I am not wrong to-day. I was right formerly, and what is taking place at present is another proof of it. Let us take, for example, the calorimetric experiment of Curie with radium. Is it possible to reconcile it with the principle of the conservation of energy? This has been attempted in many ways; but there is one among them to which I wish to call your attention; it is not the explanation which is tending to-day to prevail, but it is one of those that have been suggested. Radium is assumed to be only an intermediary, merely to store radiations of an unknown nature that fly through space in all directions, traversing all bodies except radium without being changed by this passage, and without exerting on them any action whatever. Radium alone can appropriate a little of their energy and then return it to us in various forms.

How useful this explanation is and how convenient! In the first place it is non-verifiable and hence irrefutable. Then, it can serve to account for any contradiction to Mayer's principle; it answers in advance not only the objection of Curie, but all other objections that the experimenters of the future may accumulate. This new and unknown energy could be used for anything.

That is exactly what I said, and by such means it is easy to show that our principle is safe from experimental attacks.

But then, what have we gained by this stroke? The principle is intact, but henceforth what is it good for? It enabled us to foresee that under such and such conditions we could count on a certain amount of energy; it imposed a limit; but now that there has been placed at our disposal this indefinite supply of new energy, we are no longer limited by anything; and, as I have also written, if a principle ceases to be productive, experiment, without contradicting it directly, would nevertheless condemn it.

That, then, is not what should be done. We should have to rebuild from the beginning. If we were driven to this necessity, we could easily console ourselves. We should not be obliged to conclude that science can never do aught but the work of a Penelope, that it can only raise ephemeral structures which it is soon forced to demolish completely with its own hands.

As I have said, we have already passed through a similar crisis. I have shown you that in the second mathematical physics, that of general principles, one finds traces of the first, that of central forces; it will be the same if we are to know a third. Just so with the animal that casts its outer shell, that bursts the skin that has become too small and grows a new one; under the new covering can always be recognized the essential traits of the organism that survives.

In what direction we are going to expand we are unable to foresee. Perhaps it is the kinetic theory of gases that will forge ahead and serve as a model for the others. In that case, the facts that appeared simple to us at first will be nothing more than the resultants of a very large number of elementary facts which the laws of probability alone would induce to work toward the same end. A physical law would then assume an entirely new aspect; it would no longer be merely a differential equation, it would assume the character of a statistical law.

Perhaps too we shall have to construct an entirely new mechanics, which we can only just get a glimpse of, where, the inertia increasing with the velocity, the velocity of light would be a limit beyond which it would be impossible to go. The ordinary, simpler mechanics would remain a first approximation since it would be valid for velocities that are not too great, so that the old dynamics would be found in the new. We should have no reason to regret that we believed in the older principles, and indeed since the velocities that are too great for the old formulas will always be exceptional, the safest thing to do in practice would be to act as though we continued to believe in them. They are so useful that a place should be saved for them. To wish to banish them altogether would be to deprive oneself of a valuable weapon. I hasten to say, in closing, that we are not yet at that pass, and that nothing proves as yet that they will not come out of the fray victorious and intact.

Chapter 5

Felix Klein

Adequately describing Felix Klein (1849-1925) is as difficult as reaching the bottom of his bottle.[1] His contributions extend beyond his mathematics in non- euclidean geometry, the connections between geometry and group theory, and function theory to his promotion of mathematics, his success in helping to make Göttingen Germany's main center for mathematics and the exact sciences, and his interest in teaching at all levels. Indeed, Klein was elected chairman of the International Commission on Mathematical Instruction at the Rome International Mathematical Congress of 1908. As mentioned earlier, he contributed to the development of American mathematics through the "Evanston lectures" and other activities. Excellent descriptions of him and his work can be found in Parshall and Rowe's book [4] and in Reid's [3] "Hilbert."

Already at the age of 23, Klein was a professor at Erlangen (but only for 1872-75). The first of his papers reproduced here was an insightful, seminal response to the proliferation of different geometries put forth in the first half of the nineteenth century; geometries of two, three, four and more dimensions, as well as non-Euclidean and projective geometry. Are they distinct, or are they related? This is the question Klein addressed in his inaugural lecture for his Erlangen professorship; it was published in 1872 under the title *Vergleichende Betrachtungen über neuere geometrische Forschungen*. This lecture, where Klein put forth the program of using groups to classify the various geometries, became the well known *Erlangen Program*. About twenty years later, the first English translation appeared

---

[1]A recent advertisement for a knitted "Klein bottle" cap which can be worn with a matching "Möbius strip" scarf described how this ensemble would save money; when dirty, you only need to clean one side.

in the *Bulletin of the New York Mathematics Society* [1]; it is the first article reproduced here.

As Klein writes in this *Bulletin* article, "My 1872 Programme, appearing as a separate publication ... had but a limited circulation at first. ... But now that the general development of mathematics has taken, in the meanwhile, the direction corresponding precisely to these views, and particularly since *Lie* has begun the extended form of his *Theorie der Transformations-gruppen* ... it seems proper to give a wider circulation to the expositions in my Programme." This wider exposure came through an Italian translation and the *Bulletin* article.

Around the time of transition between the eighteenth and nineteenth centuries, the research influence of Weierstrass, Dedekind, Cantor and the arguments of Klein brought to the foreground the importance of mathematical rigor in contrast to the then dominant approach of relying on intuitive proofs. Joining in this discussion, we again encounter James Pierpont [5] who argues in a May, 1899 article, "... arithmetical methods form the only sure foundation in analysis at present known." His comments dramatically underscore how the understanding of the importance of rigor along with the attitudes of that time differ from today.

> *We are all of us aware of a movement among us which Klein has so felicitously styled the arithmetization of mathematics. Few of us have much real sympathy with it, if indeed we understand it. It seems a useless waste of time to prove by laborious ϵ and δ methods what the old methods prove so satisfactorily in a few words. Indeed many of the things which exercise the mind of one whose eyes have been opened in the school of Weierstrass seem mere fads to the outsider. As well try to prove that two and two make four! ...*

> *Let me begin by remarking that there is no absolute criterion of what constitutes rigor in a demonstration. This is largely a personal matter; it also varies greatly with the period. A large part of the reasoning of the last century in analysis would probably be accepted by no one to-day as rigorous. Finally we remark that it is not necessary or even advisable to be equally rigorous at all times. In a vast and almost unexplored region where the notions dealt with are half defined and still plastic, where the methods improvised at each step are too novel to be completely tested we certainly will allow the greatest freedom to the intuition and divining power of creative effort.*

A flavor of Pierpont's article, which outlines the ideas of Klein embellished with his own enhancements, is captured by his conclusion.

> *To conclude, we believe that we have shown that the opinion the intuitionist holds in regard to the value of the evidence of our intuition is untenable. Everywhere we have seen that the notions arising from our intuition are vague and incomplete and that it is impossible to show the coextension of these notions and their arithmetical equivalents. The practice of intuitionists of supplementing their analytical reasoning at any moment by arguments drawn from intuition cannot therefore be justified.*

Pierpont's paper, which is well worth reading, and others which appeared around that time, support the historical importance of Klein's article and his arguments in advancing mathematics as we know it today. More important, however, is the original; Klein's paper [3], "The Arithmetizing of Mathematics " is reproduced here.

With his many interests, Klein viewed his major contribution to mathematics as coming from his work on function theory. The attraction seems to have been a combination of his geometric insights with Riemann's ideas and the Riemann surface; some of this work is reported in his 1882 paper "Riemanns Theorie der algebraischen Funktionen und ihre Integrale" where he treats function theory from a geometric approach. The third paper reproduced in this section [2], "Riemann and his significance for the development of modern mathematics," captures Klein's views of Riemann.

Bibliography

[1] Klein, F., A comparative review of recent researches in geometry, *Bull of New York Math Soc.* **3** (July 1893), 215-249.

[2] Klein, F., Riemann and his significance for the development of modern mathematics, *BAMS* **2** (March 1895), 165-180.

[3] Klein, F., The arithmetizing of mathematics, *BAMS* **3** (April 1896), 241-249

[4] Parshall, K. H., and D. E. Rowe, *The emergence of the American Mathematical Research Community 1876 - 1900: J. J. Sylvester, Felix Klein, and E. H. Moore*, American Mathematics Society, Providence, 1994.

[5] Pierpont, J., On the arithmetization of mathematics, *BAMS* **6** (1899), 394-406.

[6] Reid, C., *Hilbert*, Springer-Verlag, 1996.

A COMPARATIVE REVIEW OF RECENT RE-
SEARCHES IN GEOMETRY.*

*(PROGRAMME ON ENTERING THE PHILOSOPHICAL FACULTY AND
THE SENATE OF THE UNIVERSITY OF ERLANGEN IN 1872.)*

BY PROF. FELIX KLEIN.

Prefatory Note by the Author.—My 1872 Programme, appearing as a separate publication (Erlangen, A. Deichert), had but a limited circulation at first. With this I could be satisfied more easily, as the views developed in the Programme could not be expected at first to receive much attention. But now that the general development of mathematics has taken, in the meanwhile, the direction corresponding precisely to these views, and particularly since *Lie* has begun the publication in extended form of his *Theorie der Transformationsgruppen* (Leipzig, Teubner, vol. I. 1888, vol. II. 1890), it seems proper to give a wider circulation to the expositions in my Programme. An Italian translation by M. Gino Fano was recently published in the *Annali di Matematica,* ser. 2, vol. 17. A kind reception for the English translation, for which I am much indebted to Mr. Haskell, is likewise desired.

The translation is an absolutely literal one ; in the two or three places where a few words are changed, the new phrases are enclosed in square brackets []. In the same way are indicated a number of additional footnotes which it seemed desirable to append, most of them having already appeared in the Italian translation.—F. KLEIN.

Among the advances of the last fifty years in the field of geometry, the development of *projective geometry* † occupies the first place. Although it seemed at first as if the so-called metrical relations were not accessible to this treatment, as they do not remain unchanged by projection, we have nevertheless learned recently to regard them also from the projective point of view, so that the projective method now embraces the whole of geometry. But metrical properties are then to be regarded no longer as characteristics of the geometrical figures *per se*, but as their relations to a fundamental configuration, the imaginary circle at infinity common to all spheres.

When we compare the conception of geometrical figures gradually obtained in this way with the notions of ordinary (elementary) geometry, we are led to look for a general principle in accordance with which the development of both methods has been possible. This question seems the more important as, beside the elementary and the projective geometry, are arrayed a series of other methods, which, albeit they are less developed, must be allowed the same right to an individual existence. Such are the geometry of reciprocal radii

---

* Translated by Dr. M. W. HASKELL, Assistant Professor of Mathematics in the University of California.

† See Note I. of the appendix.

Reprinted from Bull. New York Math. Soc. 2 (July 1893), 215–249

vectores, the geometry of rational transformations, etc., which will be mentioned and described further on.

In undertaking in the following pages to establish such a principle, we shall hardly develop an essentially new idea, but rather formulate clearly what has already been more or less definitely conceived by many others. But it has seemed the more justifiable to publish connective observations of this kind, because geometry, which is after all one in substance, has been only too much broken up in the course of its recent rapid development into a series of almost distinct theories,* which are advancing in comparative independence of each other. At the same time I was influenced especially by a wish to present certain methods and views that have been developed in recent investigations by *Lie* and myself. Our respective investigations, different as has been the nature of the subjects treated, have led to the same generalized conception here presented; so that it has become a sort of necessity to thoroughly discuss this view and on this basis to characterize the contents and general scope of those investigations.

Though we have spoken so far only of geometrical investigations, we will include investigations on manifoldnesses of any number of dimensions,† which have been developed from geometry by making abstraction from the geometric image, which is not essential for purely mathematical investigations.‡ In the investigation of manifoldnesses the same different types occur as in geometry; and, as in geometry, the problem is to bring out what is common and what is distinctive in investigations undertaken independently of each other. Abstractly speaking, it would in what follows be sufficient to speak throughout of manifoldnesses of n dimensions simply; but it will render the exposition simpler and more intelligible to make use of the more familiar space-perceptions. In proceeding from the consideration of geometric objects and developing the general ideas by using these as an example, we follow the path which our science has taken in its development and which it is generally best to pursue in its presentation.

A preliminary exposition of the contents of the following pages is here scarcely possible, as it can hardly be presented in a more concise form; § the headings of the sections will indicate the general course of thought.

At the end I have added a series of notes, in which I have

---

* See Note II.   † See Note IV.   ‡ See Note III.

§ This very conciseness is a defect in the following presentation which I fear will render the understanding of it essentially more difficult. But the difficulty could hardly be removed except by a very much fuller exposition, in which the separate theories, here only touched upon, would have been developed at length.

either developed further single points, wherever the general exposition of the text would seem to demand it, or have tried to define with reference to related points of view the abstract mathematical one predominant in the observations of the text.

§ 1.

GROUPS OF SPACE-TRANSFORMATIONS.   PRINCIPAL GROUP. FORMULATION OF A GENERAL PROBLEM.

The most essential idea required in the following discussion is that of a *group* of space-transformations.

The combination of any number of transformations of space * is always equivalent to a single transformation. If now a given system of transformations has the property that any transformation obtained by combining any transformations of the system belongs to that system, it shall be called a *group of transformations.*†

An example of a group of transformations is afforded by the totality of motions, every motion ;being regarded as an operation performed on the whole of space. A group contained in this group is formed, say, by the rotations about one point.‡ On the other hand, a group containing the group of motions is presented by the totality of the collineations. But the totality of the dualistic transformations does not form a group; for the combination of two dualistic transformations is equivalent to a collineation. A group is, however, formed by adding the totality of the dualistic to that of the collinear transformations. §

---

* We always regard the totality of configurations in space as simultaneously affected by the transformations, and speak therefore of *transformations of space*. The transformations may introduce other elements in place of points, like dualistic transformations, for instance ; there is no distinction in the text in this regard.

† [This definition is not quite complete, for it has been tacitly assumed that the groups mentioned always include the inverse of every operation they contain ; but, when the number of operations is infinite, this is by no means a necessary consequence of the group idea, and this assumption of ours should therefore be explicitly added to the definition of this idea given in the text.]

The ideas, as well as the notation, are taken from the *theory of substitutions*, with the difference merely that there instead of the transformations of a continuous region the permutations of a finite number of discrete quantities are considered.

‡ *Camille Jordan* has formed all the groups contained in the general group of motions : *Sur les groupes de mouvements*, Annali di Matematica, vol. 2.

§ It is not at all necessary for the transformations of a group to form a continuous succession, although the groups to be mentioned in the text will indeed always have that property. For example, a group is formed by the finite series of motions which superpose a regular body upon itself, or by the infinite but discrete series which superpose a sine-curve upon itself.

Now there are space-transformations by which the geometric properties of configurations in space remain entirely unchanged. For geometric properties are, from their very idea, independent of the position occupied in space by the configuration in question, of its absolute magnitude, and finally of the sense * in which its parts are arranged. The properties of a configuration remain therefore unchanged by any motions of space, by transformation into similar configurations, by transformation into symmetrical configurations with regard to a plane (reflection), as well as by any combination of these transformations. The totality of all these transformations we designate as the *principal group* † of space-transformations; *geometric properties are not changed by the transformations of the principal group.* And, conversely, *geometric properties are characterized by their remaining invariant under the transformations of the principal group.* For, if we regard space for the moment as immovable, etc., as a rigid manifoldness, then every figure has an individual character; of all the properties possessed by it as an individual, only the properly geometric ones are preserved in the transformations of the principal group. The idea, here formulated somewhat indefinitely, will be brought out more clearly in the course of the exposition.

Let us now dispense with the concrete conception of space, which for the mathematician is not essential, and regard it only as a manifoldness of n dimensions, that is to say, of three dimensions, if we hold to the usual idea of the point as space element. By analogy with the transformations of space we speak of transformations of the manifoldness; they also form groups. But there is no longer, as there is in space, one group distinguished above the rest by its signification; each group is of equal importance with every other. As a generalization of geometry arises then the following comprehensive problem:

Given a manifoldness and a group of transformations of the same; to investigate the configurations belonging to the manifoldness with regard to such properties as are not altered by the transformations of the group.

To make use of a modern form of expression, which to be sure is ordinarily used only with reference to a particular group, the group of all the linear transformations, the problem might be stated as follows:

---

* By "sense" is to be understood that peculiarity of the arrangement of the parts of a figure which distinguishes it from the symmetrical figure (the reflected image). Thus, for example, a right-handed and a left-handed helix are of opposite "sense."

† The fact that these transformations form a group results from their very idea.

Given a manifoldness and a group of transformations of the same; to develop the theory of invariants relating to that group.

This is the general problem, and it comprehends not alone ordinary geometry, but also and in particular the more recent geometrical theories which we propose to discuss, and the different methods of treating manifoldnesses of *n* dimensions. Particular stress is to be laid upon the fact that the choice of the group of transformations to be adjoined is quite arbitrary, and that consequently all the methods of treatment satisfying our general condition are in this sense of equal value.

§ 2.

GROUPS OF TRANSFORMATIONS, ONE OF WHICH INCLUDES THE OTHER, ARE SUCCESSIVELY ADJOINED. THE DIFFERENT TYPES OF GEOMETRICAL INVESTIGATION AND THEIR RELATION TO EACH OTHER.

As the geometrical properties of configurations in space remain unaltered under *all* the transformations of the principal group, it is by the nature of the question absurd to inquire for such properties as would remain unaltered under only a part of those transformations. This inquiry becomes justified, however, as soon as we investigate the configurations of space in their relation to elements regarded as fixed. Let us, for instance, consider the configurations of space with reference to one particular point, as in spherical trigonometry. The problem then is to develop the properties remaining invariant under the transformations of the principal group, not for the configurations taken independently, but for the system consisting of these configurations together with the given point. But we can state this problem in this other form: to examine configurations in space with regard to such properties as remain unchanged by those transformations of the principal group which can still take place when the point is kept fixed. In other words, it is exactly the same thing whether we investigate the configurations of space taken in connection with the given point from the point of view of the principal group or whether, without any such connection, we replace the principal group by that partial group whose transformations leave the point in question unchanged.

This is a principle which we shall frequently apply; we will therefore at once formulate it generally, as follows:

Given a manifoldness and a group of transformations applying to it. Let it be proposed to examine the configurations contained in the manifoldness with reference to a given configuration. *We may, then, either add the given configuration to the system, and then we have to investigate the properties of the extended system from the point of view of the given group, or we may leave the system unextended, limiting the transformations to be employed to such transformations of the given group as leave the given configuration unchanged. (These transformations necessarily form a group by themselves.)*

Let us now consider the converse of the problem proposed at the beginning of this section. This is intelligible from the outset. We inquire what properties of the configurations of space remain unaltered by a group of transformations which contains the principal group as a part of itself. Every property found by an investigation of this kind is a geometric property of the configuration itself ; but the converse is not true. In the converse problem we must apply the principle just enunciated, the principal group being now the smaller. We have then :

If the principal group be replaced by a more comprehensive group, a part only of the geometric properties remain unchanged. The remainder no longer appear as properties of the configurations of space by themselves, but as properties of the system formed by adding to them some particular configuration. This latter is defined, in so far as it is a definite * configuration at all, by the following condition : The assumption that it is fixed must restrict us to those transformations of the given group which belong to the principal group.

In this theorem is to be found the peculiarity of the recent geometrical methods to be discussed here, and their relation to the elementary method. What characterizes them is just this, that they base their investigations upon an extended group of space-transformations instead of upon the principal group. Their relation to each other is defined, when one of the groups includes the other, by a corresponding theorem. The same is true of the various methods of treating manifoldnesses of n dimensions which we shall take up. We shall now consider the separate methods from this point of view, and this will afford an opportunity to explain on concrete examples the theorems enunciated in a general form in this and the preceding sections.

§ 3.

Projective Geometry.

Every space-transformation not belonging to the principal group can be used to transfer the properties of known config-

---

* Such a configuration can be generated, for instance, by applying the transformations of the principal group to any arbitrary element which cannot be converted into itself by any transformation of the given group.

urations to new ones.   Thus we apply the results of plane
geometry to the geometry of surfaces that can be represented
(*abgebildet*) upon a plane; in this way long before the origin
of a true projective geometry the properties of figures derived
by projection from a given figure were inferred from those of
the given figure.   But projective geometry only arose as it
became customary to regard the original figure as essentially
identical with all those deducible from it by projection, and
to enunciate the properties transferred in the process of projection in such a way as to put in evidence their independence
of the change due to the projection.   By this process *the
group of all the projective transformations* was made the basis
of the theory in the sense of § 1, and that is just what created
the antithesis between projective and ordinary geometry.

A course of development similar to the one here described
can be regarded as possible in the case of every kind of space-
transformation ; we shall often refer to it again.   It has gone
on still further in two directions within the domain of projective geometry itself.   On the one hand, the conception was
broadened by admitting the *dualistic* transformations into
the group of the fundamental transformations.   From the
modern point of view two reciprocal figures are not to be
regarded as two distinct figures, but as essentially one and
the same.   A further advance consisted in extending the fundamental group of collinear and dualistic transformations by
the admission in each case of the *imaginary* transformations.
This step requires that the field of true space-elements has
previously been extended so as to include imaginary elements,
—just exactly as the admission of dualistic transformations
into the fundamental group requires the simultaneous introduction of point and line as space-elements.   This is not the
place to point out the utility of introducing imaginary elements, by means of which alone we can attain an exact correspondence of the theory of space with the established system
of algebraic operations.   But, on the other hand, it must be
remembered that the reason for introducing the imaginary
elements is to be found in the consideration of algebraic operations and not in the group of projective and dualistic transformations.   For, just as we can in the latter case limit ourselves to real transformations, since the real collineations and
dualistic transformations form a group by themselves, so we
can equally well introduce imaginary space-elements even
when we are not employing the projective point of view, and
indeed must do so in strictly algebraic investigations.

How metrical properties are to be regarded from the projective point of view is determined by the general theorem of
the preceding section.   Metrical properties are to be considered as projective relations to a fundamental configuration,

the circle at infinity,* a configuration having the property
that it is transformed into itself only by those transformations
of the projective group which belong at the same time to the
principal group. The proposition thus broadly stated needs
a material modification owing to the limitation of the ordinary view taken of geometry as treating only of *real* space-
elements (and allowing only *real* transformations). In order
to conform to this point of view, it is necessary expressly to
adjoin to the circle at infinity the system of real space-elements (points); properties in the sense of elementary geometry are projectively either properties of the configurations by
themselves, or relations to this system of the real elements, or
to the circle at infinity, or finally to both.

We might here make mention further of the way in which
von Staudt in his " Geometrie der Lage " (Nürnberg, 1847)
develops projective geometry,—i.e., that projective geometry
which is based on the group containing all the real projective
and dualistic transformations.†

We know how, in his system, he selects from the ordinary
matter of geometry only such features as are preserved in
projective transformations. Should we desire to proceed to
the consideration of metrical properties also, what we should
have to do would be precisely to introduce these latter as
relations to the circle at infinity. The course of thought thus
brought to completion is in so far of great importance for the
present considerations, as a corresponding development of
geometry is possible for every one of the methods we shall
take up.

§ 4.

TRANSFER OF PROPERTIES BY REPRESENTATION (ABBILDUNG).

Before going further in the discussion of the geometrical
methods which present themselves beside the elementary and
the projective geometry, let us develop in a general form certain considerations which will continually recur in the course
of the work, and for which a sufficient number of examples
are already furnished by the subjects touched upon up to this
point. The present section and the following one will be
devoted to these discussions.

---

* This view is to be regarded as one of the most brilliant achievements of [the French school] ; for it is precisely what provides a sound
foundation for that distinction between properties of position and metrical properties, which furnishes a most desirable starting-point for projective geometry.

† The extended horizon, which includes *imaginary* transformations,
was first used by *von Staudt* as the basis of his investigations in his later
work, " Beiträge zur Geometrie der Lage " (Nürnberg, 1856–60).

Suppose a manifoldness A has been investigated with reference to a group B. If, by any transformation whatever, A be then converted into a second manifoldness A', the group B of transformations, which transformed A into itself, will become a group B', whose transformations are performed upon A'. It is then a self-evident principle that *the method of treating A with reference to B at once furnishes the method of treating A' with reference to B'*, i.e., every property of a configuration contained in A obtained by means of the group B furnishes a property of the corresponding configuration in A' to be obtained by the group B'.

For example, let A be a straight line and B the ∞^3 linear transformations which transform A into itself. The method of treating A is then just what modern algebra designates as the theory of binary forms. Now, we can establish a correspondence between the straight line and a conic section A' in the same plane by projection from a point of the latter. The linear transformations B of the straight line into itself will then become, as can easily be shown, the linear transformations B' of the conic into itself, i.e., the changes of the conic resulting from those linear transformations of the plane which transform the conic into itself.

Now, by the principle stated in § 2,* the study of the geometry of the conic section is the same, whether the conic be regarded as fixed and only those linear transformations of the plane which transform the conic into itself be taken into account, or whether all the linear transformations of the plane be considered and the conic be allowed to vary too. The properties which we recognized in systems of points on the conic are accordingly projective properties in the ordinary sense. Combining this consideration with the result just deduced, we have, then :

The theory of binary forms and the projective geometry of systems of points on a conic are one and the same, i.e., to every proposition concerning binary forms corresponds a proposition concerning such systems of points, and vice versa.†

Another suitable example to illustrate these considerations is the following. If a quadric surface be brought into correspondence with a plane by stereographic projection, the surface will have one fundamental point,—the centre of projection. In the plane there are two,—the projections of the generators passing through the centre of projection. It then follows directly: the linear transformations of the plane which

---

* The principle might be said to be applied here in a somewhat extended form.

† Instead of the plane conic we may equally well introduce a twisted cubic, or indeed a corresponding configuration in an n-dimensional manifoldness.

leave the two fundamental points unaltered are converted by the representation (*Abbildung*) into linear transformations of the quadric into itself, but only into those which leave the centre of projection unaltered.   By linear transformations of the surface into itself are here meant the changes undergone by the surface when linear space-transformations are performed which transform the surface into itself.   According to this, the projective investigation of a plane with reference to two of its points is identical with the projective investigation of a quadric surface with reference to one of its points. Now, if imaginary elements are also taken into account, the former is nothing else but the investigation of the plane from the point of view of elementary geometry.   For the principal group of plane transformations comprises precisely those linear transformations which leave two points (the circular points at infinity) unchanged.   We obtain then finally :

Elementary plane geometry and the projective investigation of a quadric surface with reference to one of its points are one and the same.

These examples may be multiplied at pleasure;* the two here developed were chosen because we shall have occasion to refer to them again.

§ 5.

On the Arbitrariness in the Choice of the Space-Element.   Hesse's Principle of Transference.   Line Geometry.

As element of the straight line, of the plane, of space, or of any manifoldness to be investigated, we may use instead of the point any configuration contained in the manifoldness,— a group of points, a curve or surface,† etc.   As there is nothing at all determined at the outset about the number of arbitrary parameters upon which these configurations shall depend, the number of dimensions of our line, plane, space, etc., may be anything we like, according to our choice of the element. *But as long as we base our geometrical investigation upon the same group of transformations, the substance of the geometry remains unchanged.*   That is to say, every proposition resulting from *one* choice of the space-element will be a true proposition under any other assumption ; but the

---

* For other examples, and particularly for the extension to higher dimensions of which those here presented are capable, let me refer to an article of mine : *Ueber Liniengeometrie und metrische Geometrie* (Mathematische Annalen, vol. 5), and further to *Lie's* investigations cited later.

† See Note III.

arrangement and correlation of the propositions will be changed.

The essential thing is, then, the group of transformations ; the number of dimensions to be assigned to a manifoldness appears of secondary importance.

The combination of this remark with the principle of the last section furnishes many interesting applications, some of which we will now develop, as these examples seem better fitted to explain the meaning of the general theory than any lengthy exposition.

Projective geometry on the straight line (the theory of binary forms) is, by the last section, equivalent to projective geometry on the conic. Let us now regard as element on the conic the point-pair instead of the point. Now, the totality of the point-pairs of the conic may be brought into correspondence with the totality of the straight lines in the plane, by letting every line correspond to that point-pair in which it intersects the conic. By this representation (*Abbildung*) the linear transformations of the conic into itself are converted into those linear transformations of the plane (regarded as made up of straight lines) which leave the conic unaltered. But whether we consider the group of the latter, or whether we base our investigation on the totality of the linear transformations of the plane, always adjoining the conic to the plane configurations under investigation, is by § 2 one and the same thing. Uniting all these considerations, we have:

The theory of binary forms and projective geometry of the plane with reference to a conic are identical.

Finally, as projective geometry of the plane with reference to a conic, by reason of the equality of its group, coincides with that projective metrical geometry which in the plane can be based upon a conic, * we can also say:

The theory of binary forms and general projective metrical geometry in the plane are one and the same.

In the preceding consideration the conic in the plane might be replaced by the twisted cubic, etc., but we will not carry this out further. The correlation here explained between the geometry of the plane, of space, or of a manifoldness of any number of dimensions is essentially identical with the principle of transference proposed by *Hesse* (Borchardt's Journal, vol. 66).

An example of much the same kind is furnished by the projective geometry of space; or, in other words, the theory of quaternary forms. If the straight line be taken as space-element and be determined, as in line geometry, by six homogeneous co-ordinates connected by a quadratic equation of

---

* See Note V.

condition, the linear and dualistic transformations of space
are seen to be those linear transformations of the six variables
(regarded as independent) which transform the equation of
condition into itself.   By a combination of considerations
similar to those just developed, we obtain the following
theorem :

*The theory of quaternary forms is equivalent to projective
measurement in a manifoldness generated by six homogeneous
variables.*

For a detailed exposition of this view I will refer to an
article in the Math. Annalen (vol. 6): "Ueber die soge-
nannte Nicht-Euklidische Geometrie" [Zweiter Aufsatz], and
to a note at the close of this paper.*

To the foregoing expositions I will append two remarks, the
first of which is to be sure implicitly contained in what has
already been said, but needs to be brought out at length,
because the subject to which it applies is only too likely to be
misunderstood.

Through the introduction of arbitrary configurations as
space-elements, space becomes of any number of dimensions
we like.   But if we then keep to the (elementary or projective) space-perception with which we are familiar, the fundamental group for the manifoldness of n dimensions is given at
the outset ; in the one case it is the principal group, in the
other the group of projective transformations.   If we wished
to take a different group as a basis, we should have to depart
from the ordinary (or from the projective) space-perception.
Thus, while it is correct to say that, with a proper choice of
space-elements, space represents manifoldnesses of any number
of dimensions, it is equally important to add that *in this
representation either a definite group must form the basis of
the investigation of the manifoldness, or else, if we wish to
choose the group, we must broaden our geometrical perception
accordingly.*   If this were overlooked, an interpretation of line
geometry, for instance, might be sought in the following way.
In line geometry the straight line has six co-ordinates; the conic
in the plane has the same number of coefficients.   The interpretation of line geometry would then be the geometry in a
system of conics separated from the aggregation of all conics
by a quadratic equation between the coefficients.   This is
correct, provided we take as fundamental group for plane
geometry the totality of the transformations represented by
the linear transformations of the coefficients of the conic
which transform the quadratic equation into itself.   But if we
retain the elementary or the projective view of plane geometry,
we have no interpretation at all.

---

* See Note VI.

The second remark has reference to the following line of reasoning : Suppose in space some group or other, the principal group for instance, be given.   Let us then select a single configuration, say a point, or a straight line, or even an ellipsoid, etc., and apply to it all the transformations of the principal group.   We thus obtain an infinite manifoldness with a number of dimensions in general equal to the number of arbitrary parameters contained in the group, but reducing in special cases, namely, when the configuration originally selected has the property of being transformed into itself by an infinite number of the transformations of the group.   Every manifoldness generated in this way may be called, with reference to the generating group, a *body*.*

If now we desire to base our investigations upon the group, selecting at the same time certain definite configurations as space-elements, and if we wish to represent uniformly things which are of like characteristics, *we must evidently choose our space-elements in such a way that their manifoldness either is itself a body or can be decomposed into bodies*.   This remark, whose correctness is evident, will find application later (§ 9).   This idea of a body will come under discussion once more in the closing section, in connection with certain related ideas.†

§ 6.

THE GEOMETRY OF RECIPROCAL RADII.   INTERPRETATION OF $x + iy$.

With this section we return to the discussion of the various lines of geometric research, which was begun in §§ 2 and 3.

As a parallel in many respects to the processes of projective geometry, we may consider a class of geometric investigations in which the transformation by reciprocal radii vectores (geometric inversion) is continually employed.   To these belong investigations on the so-called cyclides and other anallagmatic surfaces, on the general theory of orthogonal systems, likewise on potential, etc.   It is true that the processes here involved

---

* In choosing this name I follow the precedent established by *Dedekind* in the theory of numbers, where he applies the name *body* to a system of numbers formed from given elements by given operations (*Dirichlet's* Vorlesungen über Zahlentheorie, 2. Aufl.).

† [In the text sufficient attention is not paid to the fact that the proposed group may contain so-called self-conjugate subgroups.   If a geometrical configuration remain unchanged by the operations of a self-conjugate subgroup. the same is true for all configurations into which it is transformed by the operations of the whole group ; i.e., for all configurations of the body arising from it.   But a body so formed would be absolutely unsuited to represent the operations of the group. In the text, therefore, are to be admitted only bodies formed of space-elements which remain unchanged by no self-conjugate subgroup of the given group whatever.]

have not yet, like projective geometry, been united into a
special geometry, *whose fundamental group would be the totality of the transformations resulting from a combination of the
principal group with geometric inversion ;* but this may be
ascribed to the fact that the theories named have never hap-
pened to receive a connected treatment. To the individual
investigators in this line of work some such systematic conception can hardly have been foreign.

The parallel between this geometry of reciprocal radii and
projective geometry is apparent as soon as the question is
raised ; it will therefore be sufficient to call attention in a
general way to the following points :

In projective geometry the elementary ideas are the point,
line, and plane. The circle and the sphere are but special
cases of the conic section and the quadric surface. The
region at infinity of elementary geometry appears as a plane ;
the fundamental configuration to which elementary geometry
is referred is an imaginary conic at infinity.

In the geometry of reciprocal radii the elementary ideas are
the point, circle, and sphere. The line and the plane are
special cases of the latter, characterized by the property that
they contain a point which, however, has no further special
significance in the theory, namely, the point at infinity. If we
regard this point as fixed, elementary geometry is the result.

The geometry of reciprocal radii admits of being stated in
a form which places it alongside of the theory of binary forms
and of line geometry, provided the latter be treated in the
way indicated in the last section. To this end we will for the
present restrict our observations to plane geometry and
therefore to the geometry of reciprocal radii in the plane.*

We have already referred to the connection between elementary plane geometry and the projective geometry of the
quadric surface with one distinctive point (§ 4). If we disregard the distinctive point, that is to say, if we consider the
projective geometry on the surface by itself, we have a representation of the geometry of reciprocal radii in the plane. For
it is easy to see † that to the group of geometric inversion in
the plane corresponds by virtue of the representation (*Abbildung*) of the quadric surface the totality of the linear transformations of the latter into itself. We have, therefore,

The geometry of reciprocal radii in the plane and the pro-

---

* The geometry of reciprocal radii on the straight line is equivalent
to the projective investigation of the line, as the transformations in
question are the same. Thus in the geometry of reciprocal radii, also,
we can speak of the anharmonic ratio of four points on a line and of
four points on a circle.

† See the article already cited : *Ueber Liniengeometrie und metrische
Geometrie*, Mathematische Annalen, vol. 5.

jective geometry on a quadric surface are one and the same; and, similarly :

The geometry of reciprocal radii in space is equivalent to the projective treatment of a manifoldness represented by a quadratic equation between five homogeneous variables.

By means of the geometry of reciprocal radii space geometry is thus brought into exactly the same connection with a manifoldness of four dimensions as by means of [projective] geometry with a manifoldness of five dimensions.

The geometry of reciprocal radii in the plane, if we limit ourselves to *real* transformations, admits of an interesting interpretation, or application, in still another direction. For, representing the complex variable $x + iy$ in the plane in the usual way, to its linear transformations corresponds the group of geometric inversion, with the above-mentioned restriction to real operations.* But the investigation of functions of a complex variable, regarded as subject to any linear transformations whatever, is merely what, under a somewhat different mode of representation, is called the theory of binary forms. In other words :

The theory of binary forms finds interpretation in the geometry of reciprocal radii in the real plane, and precisely in the way in which complex values of the variables are represented.

From the plane we will ascend to the quadric surface, to return to the more familiar circle of ideas of the projective transformations. As we have taken into consideration only real elements of the plane, it is not a matter of indifference how the surface is chosen; it can evidently not be a ruled surface. In particular, we may regard it as a spherical surface,—as is customary for the interpretation of a complex variable,—and obtain in this way the theorem :

The theory of the binary forms of a complex variable finds representation in the projective geometry of the real spherical surface.

I could not refrain from setting forth in a note † how admirably this interpretation illustrates the theory of binary cubics and quartics.

---

* [The language of the text is inexact. To the linear transformations $z' = \dfrac{\alpha z + \beta}{\gamma z + \delta}$ (where $z' = x' + iy'$, $z = x + iy$) correspond only those operations of the group of geometric inversion by which no reversion of the angles takes place (in which the two circular points of the plane are not interchanged). If we wish to include the whole group of geometric inversion, we must, in addition to the transformations mentioned, take account of the other (not less important) ones given by the formula $z' = \dfrac{\alpha \bar{z} + \beta}{\gamma \bar{z} + \delta}$ (where again $z' = x' + iy'$, but $\bar{z} = x - iy$).]

† See Note VII.

§ 7.

EXTENSION OF THE PRECEDING CONSIDERATIONS. LIE'S SPHERE GEOMETRY.

With the theory of binary forms, the geometry of reciprocal radii, and line geometry, which in the foregoing pages appear co-ordinated and only distinguished by the number of variables, may be connected certain further developments, which shall now be explained. In the first place, these developments are intended to illustrate with new examples the idea that the group determining the treatment of given subjects can be extended indefinitely; but, in the second place, the intention was particularly to explain the relation to the views here set forth of certain considerations presented by *Lie* in a recent article.* The way by which we here arrive at *Lie's* sphere geometry differs in this respect from the one pursued by *Lie*, that he proceeds from the conceptions of line geometry, while we assume a smaller number of variables in our exposition. This will enable us to be in agreement with the usual geometric perception and to preserve the connection with what precedes. The investigation is independent of the number of variables, as *Lie* himself has already pointed out (Göttinger Nachrichten, 1871, Nos. 7, 22). It belongs to that great class of investigations concerned with the projective discussion of quadratic equations between any number of variables, —investigations upon which we have already touched several times, and which will repeatedly meet us again (see § 10, for instance).

I proceed from the connection established between the real plane and the sphere by stereographic projection. In § 5 we connected plane geometry with the geometry on a conic section by making the straight line in the plane correspond to the point-pair in which it meets the conic. Similarly we can establish a connection between space geometry and the geometry on the sphere, by letting every plane of space correspond to the circle in which it cuts the sphere. If then by stereographic projection we transfer the geometry on the sphere from the latter to the plane (every circle being thereby transformed into a circle), we have the following correspondence:

the space geometry whose element is the plane and whose group is formed of the linear transformations converting a sphere into itself, and

the plane geometry whose element is the circle and whose group is the group of geometric inversion.

---

* *Partielle Differentialgleichungen und Complexe*, Mathematische Annalen, vol. 5.

The former geometry we will now generalize in two directions by substituting for its group a more comprehensive group. The resulting extension may then be immediately transferred to plane geometry by representation (*Abbildung*).

Instead of those linear transformations of space (regarded as made up of planes) which convert the sphere into itself, it readily suggests itself to select either the totality of the *linear* transformations of space, or the totality of those plane-transformations which leave the sphere unchanged [in a sense yet to be explained]; in the former case we dispense with the sphere, in the latter with the linear character of the transformations. The former generalization is intelligible without further explanation; we will therefore consider it first and follow out its importance for plane geometry. To the second case we shall return later, and shall then in the first place have to determine the most general transformation of that kind.

Linear space-transformations have the common property of converting pencils and sheafs of planes into like pencils and sheafs. Now, transferred to the sphere, the pencil of planes gives a pencil of circles, i.e., a system of ∞^1 circles with common intersections: the sheaf of planes gives a sheaf of circles, i.e., a system of x^2 circles perpendicular to a fixed circle (the circle whose plane is the polar plane of the point common to the planes of the given sheaf). Hence to linear space-transformations there correspond on the sphere, and furthermore in the plane, circle-transformations characterized by the property that they convert pencils and sheafs of circles into the same.* *The plane geometry which employs the group of transformations thus obtained is the representation of ordinary projective space geometry.* In this geometry the point cannot be used as element of the plane, for the points do not form a *body* (§ 5) for the chosen group of transformations; but circles shall be chosen as elements.

In the case of the second extension named, the first question to be settled is with regard to the nature of the group of transformations in question. The problem is, to find plane-transformations converting every [pencil] of planes whose [axis touches] the sphere into a like [pencil]. For brevity of expression, we will first consider the reciprocal problem and, moreover, go down a step in the number of dimensions; we will therefore look for point-transformations of the plane which convert every tangent to a given conic into a like tangent. To this end we regard the plane with its conic as the representation of a quadric surface projected on the plane

---

* Such transformations are considered in Grassmann's *Ausdehnungs-lehre* (edition of 1862, p. 278).

from a point of space not in the surface in such a way that the conic in question represents the boundary curve. To the tangents to the conic correspond the generators of the surface, and the problem is reduced to that of finding the totality of the point-transformations of the surface into itself by which generators remain generators.

Now, the number of these transformations is, to be sure, ∞^n, where n may have any value. For we only need to regard the point on the surface as intersection of the generators of the two systems, and to transform each system of lines into itself in any way whatever. But among these are in particular the linear transformations, and to these alone will we attend. For, if we had to do, not with a surface, but with an n-dimensional manifoldness represented by a quadratic equation, the linear transformations alone would remain, the rest would disappear.*

These linear transformations of the surface into itself, transferred to the plane by projection (other than stereographic), give two-valued point-transformations, by which from every tangent to the boundary conic is produced, it is true, a tangent, but from every other straight line in general a conic having double contact with the boundary curve. This group of transformations will be conveniently characterized by basing a projective measurement upon the boundary conic. The transformations will then have the property of converting points whose distance apart is zero by this measurement, and also points whose distance from a given point is constant, into points having the same properties.

All these considerations may be extended to any number of variables, and can in particular be applied to the original inquiry, which had reference to the sphere and plane as elements. We can then give the result an especially perspicuous form, because the angle formed by two planes according to the projective measurement referred to a sphere is equal to the angle in the ordinary sense formed by the circles in which they intersect the sphere.

We thus obtain upon the sphere, and furthermore in the plane, a group of circle-transformations having the property that *they convert circles which are tangent to each other (include a zero angle), and also circles making equal angles with another circle, into like circles.* The group of these transformations contains on the sphere the linear transformations,

---

* If the manifoldness be stereographically projected, we obtain the well-known theorem : in regions of n dimensions (even in space) there are no isogonal point-transformations except the transformations of the group of geometric inversion. In the plane, on the other hand, there are any number besides. See the articles by *Lie* already cited.

in the plane the transformations of the group of geometric inversion.*

The circle geometry based on this group is analogous to the sphere geometry which *Lie* has devised for space and which appears of particular importance for investigations on the curvature of surfaces. It includes the geometry of reciprocal radii in the same sense as the latter includes elementary geometry.

The circle- (sphere-) transformations thus obtained have, in particular, the property of converting circles (spheres) which touch each other into circles (spheres) having the same property. If we regard all curves (surfaces) as envelopes of circles (spheres), then it results from this fact that curves (surfaces) which touch each other will always be transformed into curves (surfaces) having the same property. The transformations in question belong, therefore, to the class of *contact-transformations* to be considered from a general standpoint further on, i.e., transformations under which the contact of point-configurations is an invariant relation. The first circle-transformations mentioned in the present section, which find their parallel in corresponding sphere-transformations, are not contact-transformations.

While these two kinds of generalization have here been applied only to the geometry of reciprocal radii, they nevertheless hold in a similar way for line geometry and in general for the projective investigation of a manifoldness defined by a quadratic equation, as we have already indicated, but shall not develop further in this connection.

---

* [Perhaps the addition of some few analytical formulæ will materially help to explain the remarks in the text. Let the equation of the sphere, which we project stereographically on the plane, be in ordinary tetrahedral co-ordinates:

$$x_1^2 + x_2^2 + x_3^2 + x_4^2 = 0.$$

The x's satisfying this equation of condition we then interpret as tetracyclic co-ordinates in the plane.

$$u_1 x_1 + u_2 x_2 + u_3 x_3 + u_4 x_4 = 0$$

will be the general circular equation of the plane. If we compute the radius of the circle represented in this way, we come upon the square root $\sqrt{u_1^2 + u_2^2 + u_3^2 + u_4^2}$, which we may denote by $i u_5$. We can now regard the circles as elements of the plane. The group of geometric inversion is then represented by the totality of those homogeneous linear transformations of u_1, u_2, u_3, u_4, by which $u_1^2 + u_2^2 + u_3^2 + u_4^2$ is converted into a multiple of itself. But the extended group which corresponds to *Lie*'s sphere geometry consists of those homogeneous linear transformations of the five variables u_1, u_2, u_3, u_4, u_5, which convert $u_1^2 + u_2^2 + u_3^2 + u_4^2 + u_5^2$ into a multiple of itself.]

§ 8.

ENUMERATION OF OTHER METHODS BASED ON A GROUP
OF POINT-TRANSFORMATIONS.

Elementary geometry, the geometry of reciprocal radii, and
likewise projective geometry, if we disregard the dualistic
transformations connected with the interchange of the space-
element, are included as special cases among the large number
of conceivable methods based on groups of point-transformations.   We will here mention especially only the three following methods, which agree in this respect with those named.
Though these methods are far from having been developed
into independent theories in the same degree as projective
geometry, yet they can clearly be traced in the more recent
investigations.*

1. *The Group of Rational Transformations.*

In the case of rational transformations we must carefully
distinguish whether they are rational for *all* points of the
region under consideration, viz., of space, or of the plane, etc.,
or only for the points of a manifoldness contained in the
region, viz., a surface or curve.   The former alone are to be
employed when the problem is to develop a geometry of space
or of the plane in the meaning hitherto understood ; the latter
obtain a meaning, from our point of view, only when we wish
to study the geometry on a given surface or curve.   The same
distinction is to be drawn in the case of the *analysis situs* to
be discussed presently.

The investigations in both subjects up to this time have
been occupied mainly with transformations of the second
kind.   Since in these investigations the question has not been
with regard to the geometry on the surface or curve, but
rather to find the criteria for the transformability of two surfaces or curves into each other, they are to be excluded from
the sphere of the investigations here to be considered.†   For

---

* [Groups with a finite number of parameters having been treated
in the examples hitherto taken up, the so-called infinite groups will now
be the subject of consideration in the text.]

† [From another point of view they are brought back again, which I
did not yet know in 1872, very nicely into connection with the considerations in the text.   Given any algebraic configuration (curve, or surface, etc.), let it be transferred into a higher space by introducing the
ratios

$$\phi_1 : \phi_2 : \ldots : \phi_p$$

of the intergrands of the first species belonging to it as homogeneous
co-ordinates.   In this space we have then simply to take the group of
homogeneous linear transformations as a basis for our further considerations.   See various articles by *Brill*, *Nöther*, and *Weber*, and (to mention a single recent article) my own paper : *Zur Theorie der Abelschen
Functionen* in vol. 36 of the Math. Annalen.]

the general synopsis here outlined does not embrace the entire field of mathematical research, but only brings certain lines of thought under a common point of view.

Of such a geometry of rational transformations as must result on the basis of the transformations of the first kind, only a beginning has so far been made. In the region of the first grade, viz., on the straight line, the rational transformations are identical with the linear transformations and therefore furnish nothing new. In the plane we know the totality of rational transformations (the Cremona transformations); we know that they can be produced by a combination of quadratic transformations. We know further certain invariant properties of plane curves [with reference to the totality of rational transformations], viz., their deficiency, the existence of moduli; but these considerations have not yet been developed into a geometry of the plane, properly speaking, in the meaning here intended. In space the whole theory is still in its infancy. We know at present but few of the rational transformations, and use them to establish correspondences between known and unknown surfaces.

2. *Analysis situs.*

In the so-called analysis situs we try to find what remains unchanged under transformations resulting from a combination of infinitesimal distortions. Here, again, we must distinguish whether the whole region, all space, for instance, is to be subjected to the transformations, or only a manifoldness contained in the same, a surface. It is the transformations of the first kind on which we could found a space geometry. Their group would be entirely different in constitution from the groups heretofore considered. Embracing as it does all transformations compounded from (real) infinitesimal point-transformations, it necessarily involves the limitation to real space-elements, and belongs to the domain of arbitrary functions. This group of transformations can be extended to advantage by combining it with those real collineations which at the same time affect the region at infinity.

3. *The Group of all Point-transformations.*

While with reference to this group no surface possesses any individual characteristics, as any surface can be converted into any other by transformations of the group, the group can be employed to advantage in the investigation of higher configurations. Under the view of geometry upon which we have taken our stand, it is a matter of no importance that these configurations have hitherto not been regarded as geo-

metric, but only as analytic, configurations, admitting occasionally of geometric application, and, furthermore, that in
their investigation methods have been employed (these very
point-transformations, for instance) which we have only recently begun to consciously regard as geometric transformations. To these analytic configurations belong, above all,
homogeneous differential expressions, and also partial differential equations. For the general discussion of the latter,
however, as will be explained in detail in the next section,
the more comprehensive group of all contact-transformations
seems to be more advantageous.

The principal theorem in force in the geometry founded
on the group of all point-transformations is this: *that for an
infinitesimal portion of space a point-transformation always
has the value of a linear transformation.* Thus the developments of projective geometry will have their meaning for infinitesimals; and, whatever be the choice of the group for the
treatment of the manifoldness, *in this fact lies a distinguishing characteristic of the projective view.*

Not having spoken for some time of the relation of
methods of treatment founded on groups, one of which
includes the other, let us now give one more example of the
general theory of § 2. We will consider the question how
projective properties are to be understood from the point of
view of "all point-transformations," disregarding here the
dualistic transformations which, properly speaking, form
part of the group of projective geometry. This question is
identical with the other question, What condition differenti-
ates the group of linear point-transformations from the totality of point-transformations? What characterizes the linear
group is this, that to every plane it makes correspond a plane;
it contains those transformations under which the manifoldness of planes (or, what amounts to the same thing, of
straight lines) remains unchanged. *Projective geometry is to
be obtained from the geometry of all point-transformations by
adjoining the manifoldness of planes, just as elementary is
obtained from projective geometry by adjoining the imaginary
circle at infinity.* Thus, for instance, from the point of view
of all point-transformations the designation of a surface as an
algebraic surface of a certain order must be regarded as an
invariant relation to the manifoldness of planes. This becomes very clear if we connect, as *Grassmann* (Crelle's Journal, vol. 44) does, the generation of algebraic configurations
with their construction by lines.

§ 9.

ON THE GROUP OF ALL CONTACT-TRANSFORMATIONS.

Particular cases of contact-transformations have been long known; *Jacobi* has even made use of the most general contact-transformations in analytical investigations, but an effective geometrical interpretation has only been given them by recent researches of *Lie's*.* It will therefore not be superfluous to explain here in detail what a contact-transformation is. In this we restrict ourselves, as hitherto, to point-space with its three dimensions.

By a contact-transformation is to be understood, analytically speaking, any substitution which expresses the values of the variables x, y, z and their partial derivatives $\dfrac{dz}{dx} = p$, $\dfrac{dz}{dy} = q$ in terms of new variables x', y', z', p', q'. It is evident that such substitutions, in general, convert surfaces that are in contact into surfaces in contact, and this accounts for the name. Contact-transformations are divided into three classes (the point being taken as space-element), viz., those in which *points* correspond to the ∞^3 points (the point-transformations just considered); those converting the points into curves; lastly, those converting them into surfaces. This classification is not to be regarded as essential, inasmuch as for other ∞^3 space-elements, say for planes, while a division into three classes again occurs, it does not coincide with the division occurring under the assumption of points as elements.

If a point be subjected to all the contact-transformations it is converted into the totality of points, curves, and surfaces. Only in their entirety, then, do points, curves, and surfaces form a *body* of our group. From this may be deduced the general rule that the formal treatment of a problem from the point of view of all contact-transformations (e.g., the theory of partial differential equations considered below) must be incomplete if we operate only with point- (or plane-) co-ordinates, for the very reason that the chosen space-elements do not form a body.

If, however, we wish to preserve the connection with the ordinary methods, it will not do to introduce as space-elements all the individual configurations contained in the body, as their number is ∞^3. This makes it necessary to

---

* See, in particular, the article already cited: *Ueber partielle Differentialgleichungen und Complexe*, Mathematische Annalen, vol. 5. For the details given in the text in regard to partial differential equations I am indebted mainly to oral communications of *Lie's*; see his note, *Zur Theorie partieller Differentialgleichungen*, Göttinger Nachrichten, October 1872.

introduce in these considerations as space-element not the point, curve, or surface, but the "surface-element," i.e., the system of values x, y, z, p, q. Each contact-transformation converts every surface-element into another; the ∞^5 surface-elements accordingly form a body.

From this point of view point, curve, and surface must be uniformly regarded as aggregates of surface-elements, and indeed of ∞^2 elements. For the surface is covered by ∞^2 elements, the curve is tangent to the same number, through the point pass the same number. But these aggregates of ∞^2 elements have another characteristic property in common. Let us designate as the *united position* of two consecutive surface-elements x, y, z, p, q and $x + dx, y + dy, z + dz, p + dp, q + dq$ the relation defined by the equation

$$dz - pdx - qdy = 0.$$

Thus point, curve, and surface agree in being *manifoldnesses of ∞^2 elements, each of which is united in position with the ∞^1 adjoining elements*. This is the common characteristic of point, curve, and surface; and this must serve as the basis of the analytical investigation, if the group of contact-transformations is to be used.

The united position of consecutive elements is an invariant relation under any contact-transformation whatever. And, conversely, contact-transformations may be defined as *those substitutions of the five variables x, y, z, p, q, by which the relation*

$$dz - pdx - qdy = 0$$

is converted into itself. In these investigations space is therefore to be regarded as a manifoldness of five dimensions; and this manifoldness is to be treated by taking as fundamental group the totality of the transformations of the variables which leave a certain relation between the differentials unaltered.

First of all present themselves as subjects of investigation the manifoldnesses defined by one or more equations between the variables, i.e., *by partial differential equations of the first order, and systems of such equations.* It will be one of the principal problems to select out of the manifoldnesses of elements satisfying given equations systems of ∞^1, or of ∞^2, elements which are all united in position with a neighboring element. A question of this kind forms the sum and substance of the problem of the solution of a partial differential equation of the first order. It can be formulated in the following way: to select from among the ∞^4 elements satisfying the equation all the twofold manifoldnesses of the given kind. The problem of the complete solution thus assumes the definite form: to classify in some way the ∞^4 elements satisfying the equation into ∞^3 manifoldnesses of the given kind.

It cannot be my intention to pursue this consideration of partial differential equations further; on this point I refer to *Lie*'s articles already cited. I will only point out one thing further, that from the point of view of the contact-transformations a partial differential equation of the first order has no invariant, that every such equation can be converted into any other, and that therefore linear equations in particular have no distinctive properties. Distinctions appear only when we return to the point of view of the point-transformations.

The groups of contact-transformations, of point-transformations, finally of projective transformations, may be defined in a uniform manner which should here not be passed over.[*] Contact-transformations have already been defined as those transformations under which the united position of consecutive surface-elements is preserved. But, on the other hand, point-transformations have the characteristic property of converting consecutive line-elements which are united in position into line-elements similarly situated; and, finally, linear and dualistic transformations maintain the united position of consecutive connex-elements. By a connex-element is meant the combination of a surface-element with a line-element contained in it; consecutive connex-elements are said to be united in position when not only the point but also the line-element of one is contained in the surface-element of the other. The term connex-element (though only preliminary) has reference to the configurations recently introduced into geometry by *Clebsch*[†] and represented by an equation containing simultaneously a series of point-coordinates as well as a series of plane- and a series of line-coordinates whose analogues in the plane *Clebsch* denotes as connexes.

§ 10.

ON MANIFOLDNESSESS OF ANY NUMBER OF DIMENSIONS.

We have already repeatedly laid stress on the fact that in connecting the expositions thus far with space-perception we have only been influenced by the desire to be able to develop abstract ideas more easily through dependence on graphic

---

[*] I am indebted to a remark of *Lie*'s for these definitions.

[†] Göttinger Abhandlungen, 1872 (vol. 17): *Ueber eine Fundamentalaufgabe der Invariantentheorie*, and especially Göttinger Nachrichten, 1872, No. 22: *Ueber ein neues Grundgebilde der analytischen Geometrie der Ebene.*

examples. But the considerations are in their nature independent of the concrete image, and belong to that general field of mathematical research which is designated as the theory of manifoldnesses of any dimensions,—called by *Grassmann* briefly "theory of extension" (Ausdehnungslehre). How the transference of the preceding development from space to the simple idea of a manifoldness is to be accomplished is obvious. It may be mentioned once more in this connection that in the abstract investigation we have the advantage over geometry of being able to choose arbitrarily the fundamental group of transformations, while in geometry a minimum group—the principal group—was given at the outset.

We will here touch, and that very briefly, only on the following three methods :

1. *The Projective Method or Modern Algebra* (*Theory of Invariants*).

Its group consists of the totality of linear and dualistic transformations of the variables employed to represent individual configurations in the manifoldness; it is the generalization of projective geometry. We have already noticed the application of this method in the discussion of infinitesimals in a manifoldness of one more dimension. It includes the two other methods to be mentioned, in so far as its group includes the groups upon which those methods are based.

2. *The Manifoldness of Constant Curvature.*

The notion of such a manifoldness arose in *Riemann's* theory from the more general idea of a manifoldness in which a differential expression in the variables is given. In his theory the group consists of the totality of those transformations of the variables which leave the given expression unchanged. On the other hand, the idea of a manifoldness of constant curvature presents itself when a projective measurement is based upon a given quadratic equation between the variables. From this point of view as compared with *Riemann's* the extension arises that the variables are regarded as complex; the variability can be limited to the real domain afterwards. Under this head belong the long series of investigations touched on in §§ 5, 6, 7.

3. *The Plane Manifoldness.*

Riemann designates as a plane manifoldness one of constant zero curvature. Its theory is the immediate generalization of elementary geometry. Its group can, like the principal group of geometry, be separated from out the group of the projective method by supposing a configuration to remain fixed which is defined by two equations, a linear and a quadratic equation. We have then to distinguish between real and imaginary if we wish to adhere to the form in which the theory is usually presented. Under this head are to be counted, in the first place, elementary geometry itself, then for instance the recent generalizations of the ordinary theory of curvature, etc.

Concluding Remarks.

In conclusion we will introduce two further remarks closely related to what has thus far been presented,—one with reference to the analytic form in which the ideas developed in the preceding pages are to be represented, the other marking certain problems whose investigation would appear important and fruitful in the light of the expositions here given.

Analytic geometry has often been reproached with giving preference to arbitrary elements by the introduction of the system of co-ordinates, and this objection applies equally well to every method of treating manifoldnesses in which individual configurations are characterized by the values of variables. But while this objection has been too often justified owing to the defective way in which, particularly in former times, the method of co-ordinates was manipulated, yet it disappears when the method is rationally treated. The analytical expressions arising in the investigation of a manifoldness with reference to its group must, from their meaning, be independent of the choice of the co-ordinate system; and the problem is then to clearly set forth this independence analytically. That this can be done, and how it is to be done, is shown by modern algebra, in which the abstract idea of an invariant that we have here in view has reached its clearest expression. It possesses a general and exhaustive law for constructing invariant expressions, and operates only with such expressions. This object should be kept in view in any formal (analytical) treatment, even when other groups than the projective group form the basis of the treatment.* For the analytical formulation should, after all, be congruent with the conceptions whether it be our purpose to use it only as a precise and perspicuous expression of the conceptions, or to penetrate by its aid into still unexplored regions.

The further problems which we wished to mention arise on comparing the views here set forth with the so-called *Galois* theory of equations.

* [For instance, in the case of the groups of rotations of three-dimensional space about a fixed point, such a formalism is furnished by quaternions.]

In the *Galois* theory, as in ours, the interest centres on groups of transformations. The objects to which the transformations are applied are indeed different; there we have to do with a finite number of discrete elements, here with the infinite number of elements in a continuous manifoldness. But still the comparison may be pursued further owing to the identity of the group-idea,* and I am the more ready to point it out here, as it will enable us to characterize the position to be awarded to certain investigations begun by *Lie* and myself † in accordance with the views here developed.

In the Galois theory, as it is presented for instance in *Serret's* "Cours d'Algèbre supérieure" or in *C. Jordan's* "Traité des Substitutions," the real subject of investigation is the group theory or substitution theory itself, from which the theory of equations results as an application. Similarly we require a *theory of transformations*, a theory of the groups producible by transformations of any given characteristics. The ideas of commutativity, of similarity, etc., will find application just as in the theory of substitutions. As an application of the theory of transformations appears that treatment of a manifoldness which results from taking as a basis the groups of transformations.

In the theory of equations the first subjects to engage the attention are the symmetric functions of the coefficients, and in the next place those expressions which remain unaltered, if not under all, yet under a considerable number of permutations of the roots. In treating a manifoldness on the basis of a group our first inquiry is similarly with regard to the bodies (§ 5), viz., the configurations which remain unaltered under all the transformations of the group. But there are configurations admitting not all but some of the transformations of the group, and they are next of particular interest from the point of view of the treatment based on the group; they have distinctive characteristics. It amounts, then, to distinguishing in the sense of ordinary geometry symmetric and regular bodies, surfaces of revolution and helicoidal surfaces. If the subject be regarded from the point of view of projective geometry, and if it be further required that the transformations converting the configurations into themselves shall be commutative, we arrive at the configurations considered by *Lie* and myself in the article cited, and the general problem proposed in § 6 of that article. The deter-

---

* I should like here to call to mind *Grassmann's* comparison of combinatory analysis and extensive algebra in the introduction to the first edition of his "Ausdehnungslehre" (1844).

† See our article: *Ueber diejenigen Curven, welche durch ein geschlossenes System von einfach unendlich vielen vertauschbaren linearen Transformationen in sich übergehen*, Mathematische Annalen, Bd. IV.

mination (given in §§ 1, 3 of that article) of all the groups of an infinite number of commutative linear transformations in the plane forms a part of the general theory of transformations named above.*

NOTES.

I. *On the Antithesis between the Synthetic and the Analytic Method in Modern Geometry.*

The distinction between modern synthesis and modern analytic geometry must no longer be regarded as essential, inasmuch as both subject-matter and methods of reasoning have gradually taken a similar form in both. We choose therefore in the text as common designation of them both the term *projective geometry*. Although the synthetic method has more to do with space-perception and thereby imparts a rare charm to its first simple developments, the realm of space-perception is nevertheless not closed to the analytic method, and the formulæ of analytic geometry can be looked upon as a precise and perspicuous statement of geometrical relations. On the other hand, the advantage to original research of a well formulated analysis should not be underestimated,—an advantage due to its moving, so to speak, in advance of the thought. But it should always be insisted that a mathematical subject is not to be considered exhausted until it has become intuitively evident, and the progress made by the aid of analysis is only a first, though a very important, step.

---

* I must refrain from referring in the text to the fruitfulness of the consideration of infinitesimal transformations in the theory of differential equations. In § 7 of the article cited, *Lie* and I have shown that ordinary differential equations which admit the same infinitesimal transformations present like difficulties of integration. How the considerations are to be employed for partial differential equations, *Lie* has illustrated by various examples in several places; for instance, in the article named above (Math. Annalen, vol. 5). See in particular the proceedings of the Christiania Academy, May 1872.

[At this time I may be allowed to refer to the fact that it is exactly the two problems mentioned in the text which have influenced a large part of the further investigations of *Lie* and myself. I have already called attention to the appearance of the two first volumes of *Lie*'s "Theorie der Transformationsgruppen." Of my own work might be mentioned the later researches on regular bodies, on elliptic modular functions, and on single-valued functions with linear transformations into themselves, in general. An account of the first of these was given in a special work: "Vorlesungen über das Ikosaeder und die Auflösung der Gleichungen fünften Grades" (Leipzig, 1884); an exposition of the theory of the elliptic modular functions, elaborated by *Dr. Fricke* is in course of publication.]

II. *Division of Modern Geometry into Theories.*

When we consider, for instance, how persistently the mathematical physicist disregards the advantages afforded him in many cases by only a moderate cultivation of the projective view, and how, on the other hand, the student of projective geometry leaves untouched the rich mine of mathematical truths brought to light by the theory of the curvature of surfaces, we must regard the present state of mathematical knowledge as exceedingly incomplete and, it is to be hoped, as transitory.

III. *On the Value of Space-perception.*

When in the text we designated space-perception as something incidental, we meant this with regard to the purely mathematical contents of the ideas to be formulated. Space-perception has then only the value of illustration, which is to be estimated very highly from the pedagogical stand-point, it is true. A geometric model, for instance, is from this point of view very instructive and interesting.

But the question of the value of space-perception in itself is quite another matter. I regard it as an independent question. There is a true geometry which is not, like the investigations discussed in the text, intended to be merely an illustrative form of more abstract investigations. Its problem is to grasp the full reality of the figures of space, and to interpret—and this is the mathematical side of the question—the relations holding for them as evident results of the axioms of space-perception. A model, whether constructed and observed or only vividly imagined, is for this geometry not a means to an end, but the subject itself.

This presentation of geometry as an independent subject, apart from and independent of pure mathematics, is nothing new, of course. But it is desirable to lay stress explicitly upon this point of view once more, as modern research passes it over almost entirely. This is connected with the fact that, *vice versa*, modern research has seldom been employed in investigations on the form-relations of space-configurations, while it appears well adapted to this purpose.

IV. *On Manifoldnesses of any Number of Dimensions.*

That space, regarded as the locus of points, has only three dimensions, does not need to be discussed from the mathematical point of view; but just as little can anybody be prevented from that point of view from claiming that space really has four, or an unlimited number of dimensions, and that we are only able to perceive three. The theory of manifoldnesses, advancing as it does with the course of time more and more into the foreground of modern mathematical research, is by its nature fully independent of any such claim. But a nomenclature has become established in this theory which has indeed been derived from this idea. Instead of the elements of a manifoldness we speak of the points of a higher space, etc. The nomenclature itself has certain advantages, in that it facilitates the interpretation by calling to mind the perceptions of geometry. But it has had the unfortunate result of causing the wide-spread opinion that investigations on manifoldnesses of any number of dimensions are inseparably connected with the above-mentioned idea of the nature of space. Nothing is more unsound than this opinion. The mathematical investigations in question would, it is true, find an immediate application to geometry, if the idea were correct; but their value and purport is absolutely independent of this idea, and depends only on their own mathematical contents.

It is quite another matter when *Plücker* shows how to regard actual space as a manifoldness of any number of dimensions by introducing as space-element a configuration depending on any number of parameters, a curve, surface, etc. (see § 5 of the text).

The conception in which the element of a manifoldness (of any number of dimensions) is regarded as analogous to the point in space was first developed, I suppose, by *Grassmann* in his "Ausdehnungslehre" (1844). With him the thought is absolutely free of the above-mentioned idea of the nature of space; this idea goes back to occasional remarks by *Gauss*, and became more widely known through *Riemann's* investigations on manifoldnesses, with which it was interwoven.

Both conceptions—*Grassmann's* as well as *Plücker's*—have their own peculiar advantages; they can be alternately employed with good results.

V. *On the So-called Non-Euclidean Geometry.*

The projective metrical geometry alluded to in the text is essentially coincident, as recent investigations have shown, with the metrical geometry which can be developed under non-acceptance of the axiom of parallels, and is to-day under the name of non-Euclidean geometry widely treated and discussed. The reason why this name has not been mentioned at all in the text, is closely related to the expositions given in the preceding note. With the name non-Euclidean geometry have been associated a multitude of non-mathematical ideas, which have been as zealously cherished by some as resolutely rejected by others, but with which our purely mathematical considerations have nothing to do whatever. A wish to contribute towards clearer ideas in this matter has occasioned the following explanations.

The investigations referred to on the theory of parallels, with the results growing out of them, have a definite value for mathematics from two points of view.

They show, in the first place,—and this function of theirs may be regarded as concluded once for all,—that the axiom of parallels is not a mathematical consequence of the other axioms usually assumed, but the expression of an essentially new principle of space-perception, which has not been touched upon in the foregoing investigations. Similar investigations could and should be performed with regard to every axiom (and not alone in geometry); an insight would thus be obtained into the mutual relation of the axioms.

But, in the second place, these investigations have given us an important mathematical idea,—the idea of a manifoldness of constant curvature. This idea is very intimately connected, as has already been remarked and in § 10 of the text discussed more in detail, with the projective measurement which has arisen independently of any theory of parallels. Not only is the study of this measurement in itself of great mathematical interest, admitting of numerous applications, but it has the additional feature of including the measurement given in geometry as a special (limiting) case and of teaching us how to regard the latter from a broader point of view.

Quite independent of the views set forth is the question, what reasons support the axiom on parallels, i.e., whether we should regard it as absolutely given, as some claim, or only as approximately proved by experience, as others say. Should there be reasons for assuming the latter position, the mathematical investigations referred to afford us then immediately the means for constructing a more exact geometry. But the inquiry is evidently a philosophical one and concerns the most general foundations of our understanding. The mathematician as such is not concerned with this inquiry, and does not wish his investigations to be regarded as dependent on the answer given to the question from the one or the other point of view.*

---

* [To the explanations in the text I should like to add here two supplementary remarks.

In the first place, when I say that the mathematician as such has no stand to take on the philosophical question, I do not mean to say that the philosopher can dispense with the mathematical developments in treating the aspect of the question which interests him ; on the contrary, it is my decided conviction that a study of these developments is

VI. *Line Geometry as the Investigation of a Manifoldness
of Constant Curvature.*

In combining line geometry with the projective measurement in a manifoldness of five dimensions, we must
remember that the straight lines represent elements of the
manifoldness which, metrically speaking, are at infinity.   It
then becomes necessary to consider what the value of a system
of projective measurement is for the elements at infinity ;
and this may here be set forth somewhat at length, in order
to remove any difficulties which might else seem to stand in
the way of conceiving of line geometry as a metrical geometry.   We shall illustrate these expositions by the graphic
example of the projective measurement based on a quadric
surface.

Any two points in space have with respect to the surface
an absolute invariant,—the anharmonic ratio of the two
points together with the two points of intersection of the
line joining them with the surface.   But when the two points
move up to the surface, this anharmonic ratio becomes zero
independently of the position of the points, except in the
case where the two points fall upon a generator, when it
becomes indeterminate.   This is the only special case which
can occur in their relative position unless they coincide, and
we have therefore the theorem :

*The projective measurement in space based upon a quadric
surface does not yet furnish a measurement for the geometry
on the surface.*

This is connected with the fact that by linear transformations of the surface into itself any three points of the surface
can be brought into coincidence with three others.*

If a measurement on the surface itself be desired, we must
limit the group of transformations, and this result is obtained
by supposing any arbitrary point of space (or its polar plane)
to be fixed.   Let us first take a point not on the surface.   We
can then project the surface from the point upon a plane,
when a conic will appear as the boundary curve.   Upon this

---

the indispensable prerequisite to every philosophical discussion of the
subject.

Secondly, I have not meant to say that my *personal* interest is exhausted by the mathematical aspect of the question.   For my conception of the subject, in general, let me refer to a recent paper : "Zur
Nicht-Euklidischen Geometrie " (Math. Annalen, vol. 37).]

* These relations are different in ordinary metrical geometry; for
there it is true that two points at infinity have an absolute invariant.
The contradiction which might thus be found in the enumeration of
the linear transformations of the surface at infinity into itself is removed
by the fact that the translations and transformations of similarity contained in this group do not alter the region at infinity at all.

conic we can base a projective measurement in the plane, which must then be transferred back to the surface.* This is a measurement with constant curvature in the true sense, and we have then the theorem :

Such a measurement on the surface is obtained by keeping fixed a point not on the surface.

Correspondingly, we find : †

A measurement with zero curvature on the surface is obtained by choosing as the fixed point a point of the surface itself.

In all these measurements on the surface the generators of the surface are lines of zero length. The expression for the element of arc on the surface differs therefore only by a factor in the different cases. There is no absolute element of arc upon the surface; but we can of course speak of the angle formed by two directions on the surface.

All these theorems and considerations can now be applied immediately to line geometry. Line-space itself admits at the outset no measurement, properly speaking. A measurement is only obtained by regarding a linear complex as fixed; and the measurement is of constant or zero curvature, according as the complex is a general or a special one (a line). The selection of a particular complex carries with it further the acceptation of an absolute element of arc. Independently of this, the directions to adjoining lines cutting the given line are of zero length, and we can besides speak of the angle between any two directions.‡

VII.   *On the Interpretation of Binary Forms.*

We shall now consider the graphic illustration which can be given to the theory of invariants of binary cubics and biquadratics by taking advantage of the representation of $x + iy$ on the sphere.

A binary cubic f has a cubic covariant Q, a quadratic covariant Δ, and an invariant R.§ From f and Q a whole system of covariant sextics $Q^2 + \lambda R f^2$ may be compounded, among them being Δ^3. It can be shown‖ that every covariant of the cubic must resolve itself into such groups of six points.

---

* See § 7 of the text.
† See § 4 of the text.
‡ See the article *Ueber Liniengeometrie und metrische Geometrie*, Math. Annalen, vol. 5, p. 271.
§ See in this connection the corresponding sections of Clebsch's "Theorie der binären Formen."
‖ By considering the linear transformations of f into itself. See Math. Annalen, vol. 4, p. 352.

Inasmuch as λ can assume complex values, the number of these covariants is ∞^3.[*]

The whole system of forms thus defined can now be represented upon the sphere as follows. By a suitable linear transformation of the sphere into itself let the three points representing f be converted into three equidistant points of a great circle. Let this great circle be denoted as the equator, and let the three points f have the longitudes $0°$, $120°$, $240°$. Then Q will be represented by the points of the equator whose longitudes are $60°$, $180°$, $300°$; $\varDelta$ by the two poles. Every form $Q^2 + \lambda R f^2$ is represented by six points, whose latitude and longitude are given in the following table, where α and β are arbitrary numbers:

α	α	α	$-\alpha$	$-\alpha$	$-\alpha$
β	$120° + \beta$	$240° + \beta$	$-\beta$	$120° - \beta$	$240° - \beta$

In studying the variation of these systems of points on the sphere, it is interesting to see how they give rise to f and Q (each reckoned twice) and $\varDelta$ (reckoned three times).

A biquadratic f has a biquadratic covariant H, a sextic covariant T, and two invariants i and j. Particularly noteworthy is the pencil of biquadratic forms $iH + \lambda jf$, all belonging to the same T, among them being the three quadratic factors into which T can be resolved, each reckoned twice.

Let the centre of the sphere now be taken as the origin of a set of rectangular axes OX, OY, OZ. Their six points of intersection with the sphere make up the form T. The four points of a set $iH + \lambda jf$ are given by the following table, x, y, z being the co-ordinates of any point of the sphere :

$$
\begin{array}{rrr}
x, & y, & z, \\
x, & -y, & -z, \\
-x, & y, & -z, \\
-x, & -y, & z.
\end{array}
$$

The four points are in each case the vertices of a symmetrical tetrahedron, whose opposite edges are bisected by the co-ordinate axes ; and this indicates the rôle played by T in the theory of biquadratic equations as the resolvent of $iH + \lambda jf$.

ERLANGEN, *October*, 1872.

---

[*] [See Beltrami, *Ricerche sulla geometria delle forme binarie cubiche*, Memorie dell' Accademia di Bologna, 1870.]

THE ARITHMETIZING OF MATHEMATICS.*

An Address delivered at the public meeting of the Royal Academy of Sciences of Göttingen, November 2nd, 1895.

BY PROFESSOR FELIX KLEIN.

THOUGH the details of mathematical science, by their very nature, elude the comprehension of the layman, and therefore fail to arouse his interest, yet the mathematician may profitably indicate certain general points of view from which he surveys his science, especially if these points of view determine his attitude to kindred subjects. I propose therefore on the present occasion to explain my position in regard to an important mathematical tendency which has as its chief exponent Weierstrass, whose eightieth birthday we have lately celebrated. I refer to the *arithmetizing* of mathematics. Some account of this tendency and its origin may be given by way of preface.

The popular conception of mathematics is that of a strictly logically coördinated system, complete in itself, such as we meet with, for instance, in Euclid's geometry; but as a matter of fact, modern mathematics in its origin was of a totally different character. With the contemplation of nature as starting point, and its interpretation as object, a philosophical principle, the principle of continuity, was made fundamental; and the use of this principle characterizes the work of the great pioneers, Newton and Leibnitz, and the mathematicians of the whole of the eighteenth century—a century of discoveries in the evolution of mathematics. Gradually, however, a more critical spirit asserted itself and demanded a logical justification for the innovations made with such assurance, the establishment, as it were, of law and order after the long and victorious campaign. This was the time of Gauss and Abel, of Cauchy and Dirichlet. But this was not the end of the matter. Gauss, taking for granted the continuity of space, unhesitatingly used space intuition as a basis for his proofs; but closer investigation showed not only that many special points still needed proof, but also that space intuition had led to the too hasty assumption of the generality of certain theorems which are by no means general. Hence arose the demand for exclusively arithmetical methods of proof; nothing shall be

---

* Translated by ISABEL MADDISON, Bryn Mawr College.

Reprinted from Bull. Amer. Math. Soc. **2** (May 1896), 241–249

accepted as a part of the science unless its rigorous truth can be clearly demonstrated by the ordinary operations of analysis. A glance at the more modern text-books of the differential and integral calculus suffices to show the great change in method; where formerly a diagram served as proof, we now find continual discussions of quantities which become smaller than, or which can be taken smaller than, any given small quantity. The continuity of a variable, and what it implies, or does not imply, are discussed, and a question is brought forward whether we can, properly speaking, differentiate or integrate a function at all. This is the Weierstrassian method in mathematics, the " Weierstrass'sche Strenge," as it is called.

Of course even this assigns no absolute standard of exactness; we can introduce further refinements if still stricter limitations are placed on the association of the quantities. This is exemplified in Kronecker's refusal to employ irrational numbers, and consequent reduction of mathematics to relations between whole numbers only; and in another way in the efforts made to introduce symbols for the different logical processes, in order to get rid of the association of ideas, and the lack of accuracy which creeps in unnoticed, and therefore not allowed for, when ordinary language is used. In this connection special mention must be made of an Italian mathematician, Peano, of Turin, to whom we are indebted for various interesting notes on other points.

Summing up all these developments in the phrase, *the arithmetizing of mathematics*, I pass on to consider the influence of the tendency here described on parts of the science outside the range of analysis proper. Thus, as you see, while voluntarily acknowledging the exceptional importance of the tendency, I do not grant that the arithmetized science is the essence of mathematics; and my remarks have therefore the two-fold character of positive approbation, and negative disapproval. For since I consider that the essential point is not the mere putting of the argument into the arithmetical form, but the more rigid logic obtained by means of this form, it seems to me desirable—and this is the positive side of my thesis—to subject the remaining divisions of mathematics to a fresh investigation based on the arithmetical foundation of analysis. On the other hand I have to point out most emphatically—and this is the negative part of my task— that it is not possible to treat mathematics exhaustively by the method of logical deduction alone, but that, even at the present time, intuition has its special province. For the sake of completeness I ought also to deal with the algorithmic side of mathematics, discussing the importance of symbolic methods, but as this subject does not appeal to me personally, I shall not enter upon it. It must be understood that I have not much that is new to say on special points ; my object is rather to collect and arrange material already familiar, justifying its existence where necessary.

In the short time at my disposal I must content myself with presenting the most important points ; I begin therefore by tracing the relation of the positive part of my thesis to the domain of geometry. The arithmetizing of mathematics began originally, as I pointed out, by ousting space intuition ; the first problem that confronts us as we turn to geometry is therefore that of reconciling the results obtained by arithmetical methods with our conception of space. By this I mean that we accept the ordinary principles of analytical geometry, and try to find from these the geometrical interpretation of the more modern analytical developments. This problem, while presenting no special difficulty, has yet many ramifications, as I have had the opportunity of showing during the past year in a seminar devoted to this subject. The net result is, on the one hand, a refinement of the process of space intuition ; and on the other, an advantage due to the clearer view that is hereby obtained of the analytical results considered, with the consequent elimination of the paradoxical character that is otherwise apt to attach itself to them. What is the most general idea of a curve, of a surface? What is meant by saying of a curve, etc., that it is "analytic" or "non-analytic?" These and similar questions must be thoroughly sifted and clearly explained. The next point is that we must subject the fundamental principles of geometry to a fresh investigation. As far as the theory of the matter is concerned, this might very well be done, as it was originally, on purely geometrical lines ; but in practice on account of the overwhelming complications that present themselves, recourse must be had to the processes of analysis, that is to the methods of analytical geometry. The investigation of the formulæ by means of which we represent the different forms in space (that is, the so-called non-Euclidian geometry, and all that is connected with it) disposes of only one side, and that the more obvious one, of the inquiry ; there still remains the more important question : What justification have we for regarding the totality of points in space as a number-manifoldness in which we interpolate the irrational numbers in the usual manner between the rational numbers arranged in three directions ? We ultimately perceive that space intuition is an inexact conception, and that in order that we may subject it to mathematical treatment, we idealize it by means of the so-called axioms, which actually serve as postulates. Kerry, who died at an early age, dealt with the philosophical side of these questions, and I agree with his results in the main and especially as regards his criticism of DuBois Reymond. Conversely this fresh determination of our conception of space has in its turn given rise to new refinements of our analytical ideas. We picture before us in space an infinite number of points and forms composed of them ; from this idea have sprung the fundamental investigations on masses of points and transfinite numbers with which G. Cantor has opened up new spheres of thought to arithmetical science. Finally it is much to be desired that full use should be made of the new point of view in the further exposition of geometry, especially infinitesimal geometry ; this result will be most easily attained by treating the subject analytically. Of course, I do not mean by this a blind calculation with x, y and z, but merely a subsidiary use of these quantities when the question concerns the precise determination of boundary conditions.

From this outline of the new geometrical programme you see that it differs greatly from any that was accepted during the first half of this century, when the prevailing tendencies led to the development of projective geometry, which has long been established as a permanent constituent of our subject. Projective geometry has opened up for us with the greatest ease many new tracts of country in our science, and has been rightly called a royal road to its own particular branch of learning; our new road is on the contrary arduous and thorny, and unremitting care is needed to clear a way through the obstacles. It leads us back to what is more nearly the geometry of the ancients, and in the light of our modern ideas we learn to understand precisely the true nature of the latter, as Zeuthen has lately shown in the most brilliant manner.

Moreover we must introduce the same process of reasoning into mechanics and mathematical physics. To avoid going too much into detail I will merely explain this by two examples. Throughout applied mathematics, as in the case of space intuition, we must idealize natural objects before we can use them for purposes of mathematical argument; but we find continually that in one and the same subject we may idealize objects in different ways, according to the purpose that we have in view. To mention only a single instance, we treat matter either as continuous throughout space, or as made up of separate molecules, which we may consider to be either at rest or in rapid motion.   How and to what degree are these different hypotheses equivalent in regard to the mathematical consequences that can be deduced from them ?   The earlier expositions of Poisson and others, as also the developments of the Kinetic Theory of Gases, are not sufficiently thorough in this respect for the modern mathematician ; the problem requires to be investigated afresh ab initio.   I expect that a publication by Boltzmann, which will shortly appear, will contain some interesting conclusions on this subject.

Another question is this : Practical physics provides us plentifully with experimental results, which we unconsciously generalize and adopt as theorems about the idealized objects.   The existence of the so-called Green's function on any closed surface with an arbitrarily chosen pole, corresponding to the fact in electricity that a conductor under the influence of a charged point is in a state of electrical equilibrium, belongs to this category ; as also the theorem that every finite elastic body is capable of an infinite series of harmonic oscillations, and my deduction of the fundamental propositions of Riemann's Theory of Abelian Functions from our knowledge of the electric currents started on any conductor when the poles of a galvanic battery are applied to it.   Are these indeed, taken in the abstract, exact mathematical theorems, or how must they be limited and defined in order that they may become so?   Mathematicians have successfully sought to investigate this ; first, C. Neumann and Schwarz with their theory of Potential, and later the French school, following on the lines of the German, with the result that the theorems taken from physics have been shown to hold good to a very considerable extent.   You see here what is the precise object of these renewed investigations; not any new physical insight, but abstract mathematical argument in itself, on account of the clearness and precision which will thereby be added to our view of the experimental facts.   If I may use an expression of Jacobi's in a somewhat modified sense, it is merely a question of intellectual integrity, " die Ehre des menschlichen Geistes."

After expressing myself thus it is not easy, without running counter to the foregoing conclusions, to secure to intuition her due share in our science ; and yet it is exactly on this antithesis that the point of my present statements depends.   I am now thinking not so much of the cultivated intuition just discussed, which has been developed under the influence of logical deduction and might almost be called a form of memory; but rather of the naive intuition, largely a natural gift, which is unconsciously increased by minute study of one branch or other of the science. The word intuition (Anschauung) is perhaps not well chosen; I mean it to include that instinctive feeling for the proportion of the moving parts with which the engineer criticises the distribution of power in any piece of mechanism he has constructed; and even that indefinite conviction the practiced calculator possesses as to the convergence of any infinite process that lies before him. I maintain that mathematical intuition—so understood—is always far in advance of logical reasoning and covers a wider field.

I might now introduce a historical excursus, showing that in the development of most of the branches of our science, intuition was the starting point, while logical treatment followed. This holds in fact, not only of the origin of the infinitesimal calculus as a whole, but also of many subjects that have come into existence only in the present century. For example, I may remind you of Riemann's Theory of the Functions of a Complex Variable; and I am glad to add also that the Theory of Numbers, a subject which for a long time seemed to be most unsuited for intuitive methods of treatment, appears to have received a fresh impetus from the application of intuition in the hands of Minkowski and others. After this it would be a matter of great interest to trace from the present standpoint the development, not of particular mathematical subjects, but of the individual mathematician; but in regard to this it must suffice to mention that the two most active mathematical investigators of the present day, Lie in Leipzig, and Poincaré in Paris, both originally made use of intuitive methods. But all this, if I pursued it further, would lead us too much into detail, and finally bring us only to particular cases. I prefer to sketch the every day results of this somewhat refined intuition, as regards the quantitative, rather than the merely arithmetical or constructive, treatment of physical or technical problems. Let me refer again to the two examples from the theory of electricity already adduced; any physicist would be able to trace, without further difficulty, and with tolerable accuracy, the form of the surface of Green's function, or, in the second experiment, the shape of the lines of force in a given case. Again, consider any given differential equation, I will say, to take the most simple instance, a differential equation of the first order in two variables. Most probably the analytical method of solution fails; nevertheless we can at once find graphically the general form of the integral curves, as has recently been done for the renowned differential equation of the Problem of Three Bodies by Lord Kelvin, a master in the art of mathematical intuition. The question in all such cases, to use the language of analysis, is one of interpolation, in which less stress is laid on exactness in particular details than on a consideration of the general conditions. I will once more emphasize the fact that in stating all our laws of nature, or in trying to formulate mathematically any actual occurrence, the art lies in making a similar use of interpolation; for we have to consider the simple laws connecting the essential quantities, apart from the multitude of fortuitous disturbances. This is ultimately what I have termed above the process of idealization. Logical investigation is not in place until intuition has completed the task of idealization.

I beg that you will consider these remarks as a description, not as an explanation, of what actually occurs. The mathematician can do no more than state the character of each particular psychical operation from observations of his own mental process. Perhaps some day physiology and experimental psychology will enable us to draw more accurate conclusions as to the relation between the processes of intuition and those of logical thought. The great differences shown by observations of different individuals confirm the supposition that it is indeed a question of distinct, that is, not necessarily connected, mental activities. Modern psychologists distinguish between visual, motor and auditory endowments; mathematical intuition, as above defined, appears to belong more closely to the first two classes, and the logical method to the third class. In common with many of my fellow mathematicians I gladly welcome these investigations which psychologists have only just undertaken, for it is to be hoped that with the increase of accurate information about the psychological conditions of mathematical thought and their particular varieties, many of the differences of opinion which necessarily remain unsettled at present will disappear.

I must add a few words on mathematics from the point of view of pedagogy. We observe in Germany at the present day a very remarkable condition of affairs in this respect; two opposing currents run side by side without affecting one another appreciably. Among the teachers in our Gymnasia the need of mathematical instruction based on intuitive methods has now been so strongly and universally emphasized that one is compelled to enter a protest, and vigorously insist on the necessity for strict logical treatment. This is the central thought of a small pamphlet on elementary geometrical problems which I published last summer. Among the university professors of our subject exactly the reverse is the case; intuition is frequently not only undervalued, but as much as possible ignored. This is doubtless a consequence of the intrinsic importance of the arithmetizing tendency in modern mathematics. But the result reaches far beyond the mark. It is high time to assert openly once for all that this implies, not only a false pedagogy, but also a distorted view of the science. I gladly yield the utmost freedom to the preferences of individual academic teachers, and have always discouraged the laying-down of general rules for higher mathematical teaching, but this shall not prevent me from saying that two classes at least of mathematical lectures must be based on intuition; the elementary lectures which actually introduce the beginner to higher mathematics—for the scholar must naturally follow the same course of development on a smaller scale, that the science itself has taken on a larger—and the lectures which are intended for those whose work is largely done by intuitive methods, namely, natural scientists and engineers. Through this one-sided adherence to logical form we have lost among these classes of men much of the prestige properly belonging to mathematics, and it is a pressing and urgent duty to regain this prestige by judicious treatment.

To return to theoretical considerations, the general views which I uphold in regard to the present problems of mathematical science need scarcely be specially formulated. While I desire in every case the fullest logical working out of the material, yet I demand at the same time an intuitive grasp and investigation of the subject from all sides. Mathematical developments originating in intuition must not be considered actual constituents of the science till they have been brought into a strictly logical form. Conversely, the mere abstract statement of logical relations cannot satisfy us until the extent of their application to every branch of intuition is vividly set forth, and we recognize the manifold connections of the logical scheme, depending on the branch which we have chosen, to the other divisions of our knowledge. The science of mathematics may be compared to a tree thrusting its roots deeper and deeper into the earth and freely spreading out its shady branches to the air. Are we to consider the roots or the

branches as its essential part? Botanists tell us that the
question is badly framed, and that the life of the organism
depends on the mutual action of its different parts.

RIEMANN AND HIS SIGNIFICANCE FOR THE DE-VELOPMENT OF MODERN MATHEMATICS.*

ADDRESS DELIVERED AT THE GENERAL SESSION OF THE VER-SAMMLUNG DEUTSCHER NATURFORSCHER UND AERZTE, IN VIENNA, SEPTEMBER 27, 1894,

BY PROFESSOR FELIX KLEIN.

It is no doubt uncommonly difficult to entertain a large audience with the discussion of any mathematical question or even of the general tendencies in the development of mathematical science. This difficulty arises from the fact that the very ideas with which the mathematician works and whose multifarious connections and interrelations he investigates are the product of long-continued mental labor and are therefore far removed from the things of ordinary life.

In spite of this I did not hesitate in accepting the honor conferred upon me by the Executive Committee of your Association in requesting me to address you to-day. In doing this I was moved by the illustrious example of the great investigator, so recently deceased, who had originally been expected to speak here to you. It must always be regarded as a particular merit of Hermann von Helmholtz that, from the very beginning of his career, he took pains to present in lectures intelligible to a wider circle of scientific men the problems and results of special work in all the manifold branches of science that engaged his attention. He thus succeeded in being of assistance to each one of us in his own special field.

While for pure mathematics it would, in the nature of the case, be impossible to do this completely, it is becoming more and more recognized that in the present state of mathematical science it is eminently desirable to try, at least, to accomplish as much as can be attained in this respect. In saying this I do not express an individual opinion; I speak in the name of all the members of that Mathematical Association which was formed some years ago in connection with the Association of Naturalists and Physicians and is practically, if not formally, identical with your Section I. We cannot help feeling that in the rapid development of modern thought our science is in danger of becoming more and more isolated. The intimate mutual relation between mathematics and theoretical natural science which, to the lasting benefit of both sides, existed ever since the rise of modern analysis, threatens

---

* Translated, with the permission of the author, by ALEXANDER ZIWET.

Reprinted from Bull. Amer. Math. Soc. 1 (April 1895), 165–180

to be disrupted.   As members of the Mathematical Association we desire to counteract this serious, ever-growing danger
with all our power; it is for this reason that we decided to
meet with the general Association of Naturalists.   Through
personal intercourse with you we wish to learn how scientific
thought develops in your branches, and where accordingly
an opportunity may arise for applying the work of the
mathematician.   On the other hand, we desire and hope to
find in you a ready interest in and an intelligent appreciation
of our aims and ideas.

    With these considerations in mind I shall now attempt to
give you an idea of the life-work of BERNHARD RIEMANN, a
man who more than any other has exerted a determining influence on the development of modern mathematics.   I hope
that my remarks may prove of some interest to those, at
least, among you who are familiar with the general trend of
ideas prevailing in mechanics and theoretical physics.   But
all of you, I trust, will feel that the ideas of which I shall
have to speak form a connecting link between mathematics
and natural science.

    The outward life of Riemann may perhaps appeal to your
sympathy; but it was too uneventful to arouse particular interest.   Riemann was one of those retiring men of learning
who allow their profound thoughts to mature slowly in the
seclusion of their study.
    He was twenty-five years of age when, in 1851, he took his
doctor's degree in Göttingen, presenting a dissertation which
showed remarkable power.   It took three more years before
he became a docent at the same university.   At this time
began the appearance in rapid succession of all those important researches of which I wish to speak.   At the death of
Dirichlet, in 1859, Riemann became his successor in the University of Göttingen.   But already in 1863 he contracted
the fatal disease to which he succumbed in 1866, at the early
age of 40 years.
    His collected works, first published in 1876 by Heinrich
Weber and Dedekind, and since issued in a second edition,
are neither numerous nor extensive.   They fill an octavo
volume of 550 pages, and only about half of the matter
contained in this volume appeared during his lifetime.
The remarkable influence exerted by Riemann's work in the
past and even to the present time is entirely due to the
originality and, of course, to the *penetrating power* of his
mathematical considerations.
    While the latter characteristic cannot be the object of my
remarks to-day, I believe that you will understand the peculiar originality of Riemann's mathematical work if I point out

to you right here the unifying fundamental idea to which as a common source all his developments can be traced.

I must mention, first of all, that Riemann devoted much time and thought to physical considerations. Grown up under the great tradition which is represented by the combination of the names of Gauss and Wilhelm Weber, influenced on the other hand by Herbart's philosophy, he endeavored again and again to find a general mathematical formulation for the laws underlying all natural phenomena. These investigations do not appear to have been carried by him to a satisfactory completion; they are preserved in his posthumous papers in a very fragmentary form. We find there several incomplete attempts at a solution, all of which have, however, in common the supposition which at the present day, under the influence of Maxwell's electromagnetic theory of light, seems to be adopted by at least all younger physicists, viz., that space is filled with a continuous fluid which serves as the common medium for the propagation of optical, electrical, and gravitational phenomena.

I shall not stop to explain the details, which at the present time could only have a historical interest. The point to which I wish to call your attention is that *these physical views are the mainspring of Riemann's purely mathematical investigations*. The same tendency which in physics discards the idea of action at a distance and explains all phenomena through the internal stresses of an all-pervading ether appears in mathematics as the attempt to understand functions from the way they behave in the infinitesimal, that is, from the differential equations satisfied by them. And just as in physics any particular phenomenon depends on the general arrangement of the conditions of the experiment, thus Riemann particularizes his functions by means of the special boundary conditions which they are required to satisfy. From this point of view the formula required for the numerical calculation of the function appears as the final result, and not as the starting-point, of the investigation.

If I may be allowed to push the analogy so far, I should say that *the work of Riemann in mathematics offers a parallel to the work of Faraday in physics*. While this comparison has reference in the first place to the *qualitative* content of the leading ideas due to the two men, I believe it to hold even *quantitatively;* i.e., the results reached by Riemann are as *important* for mathematics as the results of Faraday's work are for physical science.

On the basis of this general conception let us now pass in review the various lines of Riemann's mathematical researches. It will only be natural to begin with *the theory of functions of complex variables*, which is most intimately connected with his name, although he himself may have regarded it merely as an application of tendencies having a much wider range.

The fundamental idea of this theory is well known. To investigate the functions of a variable z we substitute for this variable a binomial quantity $x + iy$ with which we operate so as to put always -1 for i^2. The result is that the properties of the ordinary functions of simple variables become intelligible to a much higher degree than would otherwise be the case. To repeat the words used by Riemann himself in his dissertation (1851) in which he laid down the fundamental ideas of his peculiar method of treating this theory: *The introduction of complex values of the variable brings out a certain harmony and regularity which otherwise would remain hidden.*

The founder of this theory is the great French mathematician Cauchy;* but only later, in Germany, did this theory assume its modern aspect which has made it the central point of our present views of mathematics. This was the result of the simultaneous efforts of two mathematicians whom we shall have to name together repeatedly,—of Riemann and Weierstrass.

While pointing to the same end, the particular methods of these two men are as different as possible; they appear almost to be opposed to each other, though, from a higher point of view, this only means that they are complementary.

Weierstrass defines the functions of a complex variable analytically by means of a common formula, viz., by infinite power-series. In all his work he avoids as far as possible the assistance to be derived from the use of geometry; his special achievement lies in the systematic rigor of his demonstrations.

Riemann, on the other hand, begins, in accordance with his general conception referred to above, with certain differential equations satisfied by the functions of $x + iy$. Thus the problem at once assumes a physical form.

Let us put $f(x + iy) = u + iv$. Each of the two component parts of the function, u as well as v, then appears, owing to the differential equations, as a *potential* in the space of the two variables x and y; and Riemann's method can be briefly characterized by saying that *he applies to these parts u and v the principles of the theory of the potential.* In other words, his starting point lies in the domain of mathematical physics.

---

* In the text I refrained from mentioning Gauss, who, being in advance of his time in this as in other fields, anticipated many discoveries without publishing what he had found. It is very remarkable that in the papers of Gauss we find occasional glimpses of methods in the theory of functions which are completely in line with the later methods of Riemann, as if unconsciously a transfer of leading ideas had taken place from the older to the younger mathematician.

You will notice that even in mathematics free play is given to individuality of treatment.

It should also be observed that the theory of the potential, which in our day, owing to its importance in the theory of electricity and in other branches of physics, is quite universally known and used as an indispensable instrument of research, was at that time in its infancy. It is true that Green had written his fundamental memoir as early as in 1828; but this paper remained for a long time almost unnoticed. In 1839 Gauss followed with his researches. As far as Germany is concerned, it is mainly due to the lectures of Dirichlet that the theory was farther developed and became known more generally; and this is where Riemann finds his base of operations.

Riemann's specific achievement in this connection consists, of course, in the first place in the tendency to make the theory of the potential of fundamental importance for the whole of mathematics; and secondly, in a series of *geometric constructions*, or, as I should prefer to say, of *geometric inventions*, of which you must allow me to say a few words.

As a first step Riemann considers the equation $u + iv = f(x + iy)$ throughout as a *representation* (Abbildung) of the xy-plane on the uv-plane. It appears that this representation is conformal, i.e., that it preserves the magnitude of the angles; indeed, it is directly characterized by this property. We have thus a new means of defining the functions of $x + iy$. Riemann develops in this way the elegant proposition that there always exists a function f that transforms any simply-connected region of the xy-plane into any given simply-connected region of the uv-plane; this function is fully determined, apart from three constants which remain arbitrary.

Next he introduces as a fundamental idea the conception of what is now called a *Riemann surface*, i.e., a surface which overlies the plane several times, the different sheets hanging together at the so-called branch-points. This step in the development, while it must have been the most difficult to take, proved at the same time of the greatest consequence. Even at the present time we can daily notice how hard it is for the beginner to understand the essential idea of the Riemann surface and how he comes at once into the possession of the whole theory as soon as he has fully grasped this fundamental mode of conceiving the function. The Riemann surface furnishes us a means of understanding many-valued functions of $x + iy$ in the course of their variation. For on this surface there exist potentials just like those on the simple plane, and their laws can be investigated by the same methods; moreover, the representation (Abbildung) is again conformal.

As the primary principle of classification appears here the order of connectivity of these surfaces, i.e., the number of the cross-sections or cuts that can be made without resolving the surface into separate portions. This is again an entirely new geometrical method of attack, which in spite of its elementary character had never been touched upon by anybody before Riemann.

Perhaps I have gone too far into the details with what I have said. I wish only to add that all these new tools and methods, created by Riemann for the purposes of pure mathematics out of the physical intuition, have again proved of the greatest value for mathematical physics. Thus, for instance, we now always make use of Riemann's methods in treating the *stationary* flow of a fluid within a two-dimensional region. A whole series of most interesting problems, formerly regarded as insolvable, has thus been solved completely. One of the best known problems of this kind is Helmholtz's determination of the shape of a free liquid jet.

Perhaps less attention has been paid to another physical application in which Riemann's ways of looking at things are laid under contribution in a most attractive manner. I have in mind the theory of *minimum surfaces*. Riemann's own investigations on this subject were not published till after his death, in 1867, almost simultaneously with the parallel investigations of Weierstrass concerning the same question. Since that time the theory has been developed much farther by Schwarz and others. The problem is to determine the shape of the least surface that can be spread out in a rigid frame,— let us say, the form of equilibrium of a fluid lamina that fits in a given contour. It is noteworthy that, with the aid of Riemann's methods, the known functions of analysis are just sufficient to dispose of the more simple cases.

These applications to which I here call particular attention represent, of course, merely one side of the matter. There can be no doubt that the main value of these new methods in the theory of functions is to be found in their use in *pure mathematics*. I must try to show this more distinctly, in accordance with the importance of the subject, without presupposing much special knowledge.

Let me begin with the very general question of the present state of progress in the domain of pure mathematics. To the layman the advance of mathematical science may perhaps appear as something purely arbitrary because the concentration on a definite given object is wanting. Still there exists a regulating influence, well recognized in other branches of science, though in a more limited degree; it is the *historical continuity. Pure mathematics grows as old problems are worked out by means of new methods. In proportion as a better understanding is thus gained for the older questions, new problems must naturally arise.*

Guided by this principle we must now briefly pass in review the working material that Riemann found ready for use in the theory of functions when he entered upon his scientific career. It had then been recognized that particular importance attaches to three classes among the various kinds of analytic functions (i.e., functions of $x + iy$). The first class comprises the *algebraic* functions which are defined by a finite number of elementary operations (addition, multiplication, and division); in contradistinction to these we have the transcendental functions whose definition requires an infinite series of such operations. Among these transcendental functions the simplest are, on the one hand, the logarithms, on the other the trigonometric functions, such as the sine, cosine, etc.

But in Riemann's time mathematical science had already advanced beyond these elementary functions, first, to the *elliptic* functions derived from the inversion of the elliptic integrals, and, second, to the *functions connected with Gauss's hypergeometric series,* viz., spherical harmonics, Bessel functions, gamma-functions, etc.

Now what Riemann accomplished may be stated briefly by saying that for each of these three classes of functions he found not only new results, but entirely new points of view which have formed a continual source of inspiration up to the present time. A few additional remarks may serve to explain this more in detail.

The study of *algebraic functions* practically coincides with the study of algebraic *curves* whose properties are investigated by the geometer, whether he calls himself an "analyst," regarding the analytical formula as of primary importance, or a "synthetic geometer," as the term was understood by Steiner and von Staudt, who operates with the row of points and the pencil of rays. The essentially new point of view, here introduced by Riemann, is that of the general single-valued transformation (or one-to-one correspondence). This point of view allows to group the innumerable variety of algebraic curves into large classes; and by making abstraction of the peculiarities of shape of the individual curves, those general properties can be studied that belong in common to all the curves of the same class. The geometers have not been slow to derive the results so obtained from their point of view and to develop them still farther; Clebsch, in particular, worked in this direction, and he even began to attack the corresponding problems for algebraic configurations of more dimensions. But it must be insisted upon that the theory of curves should try to assimilate the *methods* of Riemann in their true essential character. A first step will consist in constructing on the curve itself the analogue of the two-dimensional Riemann surface; and this can be done in various ways. A further step would have to show us how to operate with the methods of the theory of functions in the configuration thus defined.

The theory of *elliptic integrals* finds its further development in the consideration of the general integrals of algebraic functions, a subject on which the Norwegian mathematician Abel published the first fundamental investigations in the twenties of the present century. It must always be regarded as one of the greatest achievements of Jacobi that, by a sort of inspiration, he established for these integrals a problem of inversion which furnishes single-valued functions just as the simple inversion does in the case of the elliptic integrals. The actual solution of this problem of inversion is the central task performed at the same time, but by different methods, by Weierstrass and Riemann. The great memoir on the Abelian functions in which Riemann published his theory in 1857 has always been recognized as the most brilliant of all the achievements of his genius. Indeed, the result is here reached, not by laborious calculations, but in the most direct way, by a proper combination of the geometrical considerations just referred to. I have shown in another place that his results concerning the integrals, as well as the conclusions that follow for the algebraic functions, can be obtained in a very graphic manner by considering the stationary flow of a fluid, say of electricity, on closed surfaces situated in any way in space. This, however, has reference only to one half of Riemann's memoir. The second half, which is concerned with the thetaseries, is perhaps still more remarkable. The important result is here deduced that the theta-series required for the solution of Jacobi's problem of inversion are not the general thetaseries; and this leads to the new problem of determining the position of the general theta-series in our theory. According to an observation made by Hermite, Riemann must have known the proposition published at a later time by Weierstrass and recently discussed by Picard and Poincaré, that the theta-series are sufficient for defining the most general periodic functions of several variables.

But I must not enter into these details. It is difficult to give a connected account of the further development of Riemann's theory of Abelian functions because the far-reaching investigations of Weierstrass on the same subject are as yet known only from written lectures of his students. I will therefore only mention that the important treatise published by Clebsch and Gordan in 1866 had in the main the object of deriving Riemann's results on the algebraic curve by means of analytic geometry. At that time Riemann's methods were still a sort of secret science confined to his immediate pupils and were regarded almost with distrust by other mathematicians. I can in this respect only repeat what I just remarked in speaking of the theory of curves, viz., that the growing development of mathematics leads with necessity to the incorporation of Riemann's methods into the general body of mathematical science. It is interesting to compare in this respect the latest French text-books.*

The third class of functions referred to above comprises those laws of dependence that are connected with Gauss's *hypergeometric series.* In a wider sense, we have here the functions defined by linear differential equations with algebraic coefficients. Riemann published on this subject during his lifetime only a preliminary study (in 1856) which is devoted entirely to the hypergeometric case and shows in a surprising way how all the remarkable properties of the hypergeometric function, that had been known before, can be derived without calculation from the behavior of this function when the variable passes around the singular points. We now know from his posthumous papers in what form he had intended to carry out the corresponding general theory of the linear differential equations of the nth order. Here, also, he wanted to take as a starting point and primary characteristic for the classification the group of those linear substitutions which the integrals undergo when the variable passes around the singular points.

This idea, which in a certain sense corresponds to Riemann's treatment of the Abelian integrals, has not yet been carried out according to Riemann's extensive plan. The numerous investigations on linear differential equations published during the last decades have really disposed only of certain parts of the theory. In this respect the researches of Fuchs deserve special mention.

As regards the differential equations of the *second* order, they are capable of a simple geometrical illustration. It is only necessary to consider the conformal representation which the quotient of two particular integrals of the differential equation furnishes for the region of the independent variable. In the simplest case, i.e. that of the hypergeometric function, we obtain the representation of the half-plane on a triangle formed by circular arcs; and this establishes a noteworthy connection with spherical trigonometry. In the general theory there are cases which admit of one-valued (*eindeutig*) inversion and thus produce those remarkable functions of a single

---

* See, for instance, PICARD, *Traité d'analyse*, and APPELL et GOURSAT, *Théorie des fonctions algébriques et de leurs intégrales.*

variable which, like the periodic functions, are transformed into themselves by an infinite number of linear transformations and which I have therefore called *automorphic* functions.

All these developments which occupy the mathematicians of our time appear more or less explicitly in Riemann's posthumous papers, particularly in the memoir on minimum surfaces referred to above.   For further details I must refer you to Schwarz's memoir on the hypergeometric series and to the epoch-making researches of Poincaré in the theory of automorphic functions.   With this group of investigations must also be classed those on the elliptic modular functions and the functions of the regular bodies.

I cannot leave the discussion of Riemann's work in the theory of functions without mentioning an isolated paper in which the author makes interesting contributions to the theory of definite integrals.   This paper has become celebrated mainly owing to the application which the author makes of his results to a problem in the theory of numbers, viz., *the law of distribution of the prime numbers* in the natural series of numbers.   Riemann arrives at an approximate expression for this law which agrees more closely with the results obtained by actual enumerations than the empirical rules that had been deduced up to that time from such enumerations.

Two remarks naturally present themselves in this connection.   First, I should like to call your attention to the curious way in which the various branches of higher mathematics are interwoven: a problem apparently belonging to the elements of the theory of numbers is here in a most unexpected manner brought nearer to its solution by means of considerations derived from the most intricate questions in the theory of functions.   Second, I must observe that the proofs in Riemann's paper, as he notices himself, are not quite complete; in spite of numerous attempts in recent times, it has not yet been possible to make all these proofs perfectly satisfactory.   It appears that Riemann must have worked very largely by intuition.

This, by the way, is also true of his manner of establishing the foundations of the theory of functions.   Riemann here makes use of a mode of reasoning often employed in mathematical physics; he designated this method as *Dirichlet's principle,* in honor of his teacher Lejeune Dirichlet.   When it is required to determine a continuous function that makes a certain double integral a minimum, this principle asserts that the *existence* of such a function is evident from the existence of the problem itself.*   Weierstrass has shown that

* It appears from the context that, contrary to a common way of

this inference is faulty: it may happen that the required minimum represents a limit which cannot be reached within the region of the continuous functions. This consideration affects the validity of a large portion of Riemann's developments.

Nevertheless the far-reaching results based by Riemann on this principle are all correct, as has been shown later by Carl Neumann and Schwarz with the aid of rigorous methods. It must be supposed that Riemann originally took his theorems from the physical intuition which here again proved to be of the greatest value as a guide to discovery, and that he connected them afterwards with the method of inference mentioned above, in order to obtain a connected chain of mathematical reasoning. It appears from the long developments of his dissertation that in doing this he became conscious of certain difficulties; but seeing that this mode of reasoning was used without hesitation in analogous cases by those around him, even by Gauss, he does not seem to have pursued these difficulties as far as might have been desired.

So much about the functions of complex variables. They form the only branch of mathematics that Riemann has treated as a connected whole; all his other works are separate investigations of particular questions. Still we should obtain a very inadequate picture of the mathematician Riemann if we were to disregard these latter researches. For, apart from the notable results reached by him, the consideration of these researches will place in better perspective the general conception that ruled his thoughts and the programme of work that he had laid out for himself. Besides, every one of these investigations has exerted a highly stimulating and determining influence on the further development of mathematical science.

To begin with, let me state more fully what I indicated above, viz., that Riemann's treatment of the theory of functions of complex variables, founded on the partial differential equation of the potential, was intended by him to serve merely as an *example* of the analogous treatment of all other physical problems that lead to partial differential equations, or to differential equations in general. In every such case it should be inquired what discontinuities are compatible with the differential equations, and how far the solutions may be

---

speaking, I mean by the " principle " the mode of reasoning, and not the results deduced thereby.

I wish to use this opportunity to call attention to an article by Sir Wm. Thomson, published in Liouville's *Journal*, vol. 12 (1847), which does not seem to have been sufficiently noticed by German mathematicians. In this article the principle is expressed in a very general form.

determined by the discontinuities and accessory conditions occurring in the problem. The execution of this programme which has since been considerably advanced in various directions, and which has in recent years been taken up with particular success by French geometers, amounts to nothing short of a *systematic reconstruction of the methods of integration required in mechanics and mathematical physics.*

Riemann himself has treated on this basis only a single problem in greater detail, viz., the propagation of plane waves of air of finite amplitude (1860).

Two main types must be distinguished among the linear partial differential equations of mathematical physics: the elliptic and the hyperbolic type, the simplest examples being the differential equation of the potential and that of the vibrating string, respectively. As an intermediate limiting case we may distinguish the parabolic type to which belongs the differential equation of the flow of heat. The recent investigations of Picard have shown that the methods of integration used in the theory of the potential can be extended with slight modifications to the elliptic differential equations generally. How about the other types? Riemann makes in his paper an important contribution to the solution of this question. He shows what modifications must be made in the well-known "boundary problem" of the theory of the potential and in its solution by means of Green's function in order to make this method applicable to the hyperbolic differential equations.

This paper of Riemann's is noteworthy for various other reasons. Thus, the reduction of the problem named in the title to a linear differential equation is in itself no mean achievement. Another point to which I desire to call your particular attention is the *graphical treatment of the problem* which evidently underlies the whole memoir. While this mode of treatment will have nothing surprising to the physicist, its value is in our day not infrequently underrated by the mathematician accustomed to more abstract methods. I take, therefore, particular pleasure in pointing to the fact that an authority like Riemann makes use of this mode of treatment and derives by its means most interesting results.

It remains to discuss the two great memoirs presented by Riemann, at the age of 28 years, when he became a docent at Göttingen, in 1854, viz., the essay *On the hypotheses that lie at the foundation of geometry,* and the memoir *On the possibility of representing a function by means of a trigonometric series.* It is remarkable how differently these two papers have been treated by the wider scientific public. The importance of the disquisition on the hypotheses of geometry has long since been adequately recognized, no doubt mainly owing to the part that Helmholtz took in the discussion of the problem, as is probably known to most of you. The investigation of trigonometric series, on the other hand, has so far not become known outside the circle of mathematical readers. Nevertheless the results contained in this latter paper, or rather the considerations to which it has given rise and with which its subject is intimately connected, must be regarded as of the highest interest from the point of view of the theory of knowledge.

As regards the *hypotheses of geometry*, I shall not enter here upon the discussion of the philosophical significance of the subject, as I should not have anything new to add. For the mathematician the interest of the discussion centres not so much in the origin of the geometrical axioms as in their mutual logical interdependence. The most celebrated question is that as to the nature of the axiom of parallels. It is well known that the investigations of Gauss, Lobachevsky, and Bolyai (to mention only the most prominent names) have shown that the axiom of parallels is certainly not a consequence of the other axioms, and that by disregarding the axiom of parallels a more general, perfectly consistent geometry can be constructed which contains the ordinary geometry as a particular case. Riemann gave to these important considerations a new turn by introducing and applying the mode of treatment of *analytic* geometry. He regards space as a particular case of a triply-extended numerical manifoldness in which the square of the element of arc is expressed as a quadratic form of the differentials of the co-ordinates. Without discussing the special geometrical results thus obtained and the subsequent development of this theory, I only wish to point out that here again Riemann remains faithful to his fundamental idea: *to interpret the properties of things from their behavior in the infinitesimal.* He has thereby laid the foundation for a new chapter of the differential calculus, viz., *the theory of the quadratic differential expressions of any number of variables,* and in particular of the *invariants* of such expressions under any transformations of the variables. I must here depart from the prevailing character of my remarks and call special attention to the abstract side of the matter. There can be no doubt that in trying to *discover* mathematical relations it is by no means indifferent whether we endow the symbols with which we operate with a definite meaning or not; for it is just through these concrete representations that we form those associations of ideas which lead us onward. The best proof of this will be found in almost everything that I have said to-day as to the intimate relation of Riemann's mathematics to mathematical physics. But the final result of the mathematical investigation is quite independent of these considerations and rises above these special auxiliary methods; it represents a general logical framework whose content is indifferent and can be selected in various ways according to the nature of the case. Considered from this point of view it will no longer appear surprising that at a later period (1861), in the prize essay presented to the Paris Academy, Riemann made an application of his investigation of differential expressions to the problem of the flow of heat, i.e., to a subject which surely has nothing to do with the hypotheses of geometry. In a similar way these researches of Riemann are connected with the recent investigations concerning the equivalence and classification of the general problems of mechanics. For it is possible, according to Lagrange and Jacobi, to represent the differential equations of mechanics in such a way as to make them depend on a single quadratic form of the differentials of the co-ordinates.

I now pass to the consideration of the memoir *On the trigonometric series*, which I have intentionally left to the last, because it brings into prominence a final essential characteristic of Riemann's conception. In all the preceding remarks it has been possible for me to appeal to the current ideas of physics, or at least of geometry. But Riemann's penetrating mind was not satisfied with making use of the geometrical and physical intuition; he went so far as to investigate this intuition critically, and to inquire into the *necessity* of the mathematical relations flowing from it. The question at issue is nothing less than the *fundamental principles of the infinitesimal calculus*. In his other works Riemann has nowhere expressed any definite opinions concerning these questions. It is different in the paper on trigonometric series. Unfortunately he considers only detached problems, viz., the questions whether a function can be discontinuous at every point, and whether in the case of functions of such a general nature it is still possible to speak of integration. But these problems he treats in so convincing a manner that the investigations of others on the foundations of analysis have received from him a most powerful impulse.

Tradition has it that in later years Riemann pointed out to his students the fact which must be regarded as the most remarkable result of the modern critical spirit: the existence of functions which are not differentiable at any point. A more detailed study of such "nonsensical" functions (as they used to be called formerly) has, however, been made only by Weierstrass, who has probably contributed most to give its present rigorous form to the *theory of the real functions of real variables*, as this whole field is now usually designated.

As I understand Riemann's developments on trigonometric series, he would, as far as the foundation is concerned, agree with the presentation given by Weierstrass, which discards all space-intuition and operates exclusively with arithmetical definitions. But I could never believe that Riemann should have considered this space-intuition (as is now occasionally done by extreme representatives of the modern school) as something antagonistic to mathematics, which must necessarily involve faulty reasoning. I must insist on the position that a compromise is possible in this difficulty.

We touch here upon a question which I am inclined to consider as of decisive importance for the further development of mathematical science in our time. Our students are at present introduced at the very beginning to all those intricate relations whose possibility has been discovered by modern analysis. This is no doubt very desirable; but it carries with it the dangerous consequence that our young mathematicians frequently hesitate to formulate any general propositions, that they are lacking in that freshness of thought without which no success is possible in science as well as elsewhere.

On the other hand, the majority of those engaged in applied science believe that they may entirely leave aside these difficult investigations. Thus they detach themselves from rigorous science and develop for their private use a special kind of mathematics which grows up like a wild sprig from the root of the grafted tree. We must try with all our might to overcome this dangerous split between pure and applied mathematics. I may therefore be allowed to formulate my personal position concerning this matter in the following two propositions:

First, I believe that those defects of space-intuition by reason of which it is objected to by mathematicians are merely temporary, and that this intuition can be so trained that with its aid the abstract developments of the analysts can be understood, at least in their general *tendency*.

Second, I am of the opinion that, with this more highly-developed intuition, the applications of mathematics to the phenomena of the outside world will, on the whole, remain unchanged, provided we agree to regard them throughout as a sort of *interpolation* which represents things with an approximation, limited, to be sure, but still sufficient for all practical purposes.

With these remarks I will close my address, which I hope has not taxed your indulgence unduly. You may have noticed that even in mathematics there is no standstill, that the same activity prevails there as in the natural sciences. And this, too, is a general law: that while many workers contribute to the development of science, the really new impulses can be traced back to but a small number of eminent men. But the work of these men is by no means confined to the short span of their life; their influence continues to grow in proportion as their ideas become better understood in the course of time. This is certainly the case with Riemann. For this reason you must consider my remarks not as the description of a past epoch, whose memory we cherish with a feeling of veneration, but as the picture of live issues which are still at work in the mathematics of our time.

Chapter 6

David Hilbert

It is a rare graduate student in mathematics whose office, or at least the office next door, isn't adorned with a poster of David Hilbert (1861-1943), and who cannot repeat several antidotes (some may even be accurate) about him. A highly readable and entertaining history of this dominating force in twentieth century mathematics, and his interactions with other mathematicians, is Constance Reid's [3] 1969 book *Hilbert*.

And dominating he was. An interesting story demonstrating his impact involves the formation of the International Mathematical Union at the 1920 International Congress of Mathematicians. This event took place right after World War I when war sentiments were still running high. The result is that mathematicians from the "central powers" of Germany and its allies were not invited to attend this meeting, or their national societies to join the IMU. As mathematics has no political boundaries, this policy hurt the advancement of international mathematics. In reaction to this exclusionary IMU policy, the AMS withdrew support for having the next International Congress of Mathematicians in the New York; it was held in Toronto (Riehm [4]) instead. The policy finally was changed for the 1928 ICM meeting in Italy, and, as Reid (p. 188) reports,

> "... although suffering from a recurrence of his illness, Hilbert personally led a delegation of 67 mathematicians to the Congress. At the opening session, as the Germans came into an international meeting for the first time since the war, the delegates saw a familiar figure, more frail then they remembered, marching at their head. For a few minutes there was not a sound in the hall. Then, spontaneously, every person rose and applauded."

Reihm [4] continues with Hilbert's reaction, which is as valid now as then, that

> *"All limits, especially national ones, are contrary to the nature of mathematics."*

Hilbert's eclectic interests included his interest in the foundations of geometry. When his lectures on this topic appeared in the book "Foundations of Geometry, it attracted strong reviews across the mathematical world. Sommer's [5] *Bulletin* review provides the background.

> *"The committee in charge of the unveiling of the Gauss-Weber monument at Göttingen has published a memorial volume intended to commemorate the celebration and to serve as a worthy tribute to the genius of the two great men of science. Two professors of the University of Göttingen present in this volume their investigations concerning the foundations of the exact sciences; Professor Hilbert treats ... the foundations of geometry; ..."*

Sommer explains that it is Hilbert's "... aim to lay the proper foundations for euclidean geometry, and beyond this, for analytic geometry." Sommer's description in his review provides a fairly complete picture of Hilbert's presentation. Of importance to the mathematical community of the time, and to history, is the reaction of the then most dominating mathematician, Poincaré. The translated version of Poincaré's review [2] of Hilbert's book appeared in the 1903 *Bulletin*, and it is reproduced here.

The connections between the best mathematician of the time, Poincaré, and the heir apparent, Hilbert, has several interesting connections including the influential Hilbert's problems. When Hilbert was invited to the 1900 ICM, he wanted to give a talk that would be appropriate for the beginning of the new century; but what should he talk on? While Hilbert was impressed by Poincaré's presentation on the reciprocal relationship between physics and mathematics at the previous ICM (which is reproduced in the chapter on Poincaré),

> *"He had always wanted to reply to Poincaré with a defense of mathematics for its own sake, but he had another idea. He had frequently reflected upon the importance of individual problems in the development of mathematics. Perhaps he could discuss the direction of mathematics in the coming century in terms of certain important problems on which mathematicians should concentrate their efforts."* (Reid [3], p. 68)

The first English translation of Hilbert's talk appeared in the *Bulletin*; it is reproduced here. What a delight to read; full of intuition along with advice that, while perhaps more difficult to realize today, remains valid; e.g., "An old French mathematician said: 'A mathematical theory is not to be considered complete until you have made it so clear that you can explain it to the first man whom you meet on the street.' "

Bibliography

[1] Hilbert, D., Mathematical Problems, *BAMS* **8** (1902), 437-479.

[2] Poincaré, H., Review of Hilbert's "Foundations of Geometry" *BAMS* **10** (1903), 1-23.

[3] Reid, C., *Hilbert*, Springer-Verlag, 1996.

[4] Riehm, C., The Early History of the Field's Medal, *Notices of the American Mathematical Society* **49** (2002), 778-782

[5] Sommer, J., Hilbert's Foundations of Geometry, *BAMS* **7** (1900), 287-299

BULLETIN OF THE

AMERICAN MATHEMATICAL SOCIETY.

POINCARÉ'S REVIEW OF HILBERT'S "FOUNDATIONS OF GEOMETRY." *

What are the fundamental principles of geometry? what is its origin? its nature? its scope? These are questions which have at all times engaged the attention of mathematicians and thinkers, but which about a century ago took on an entirely new aspect, thanks to the ideas of Lobachevsky and of Bolyai.

For a long time we attempted to demonstrate the proposition known as the *postulate of Euclid;* we constantly failed; we know now the reason for these failures. Lobachevsky succeeded in building a logical edifice as coherent as the geometry of Euclid, but in which the famous postulate is assumed false, and in which the sum of the angles of a triangle is always less than two right angles. Riemann devised another logical system, equally free from contradiction, in which this sum is on the other hand always greater than two right angles. These two geometries, that of Lobachevsky and that of Riemann, are what are called the *non-euclidean geometries.* The postulate

---

* Translated for the BULLETIN, with the author's permission, by Dr. E. V. HUNTINGTON. The original review appeared in Darboux's *Bulletin des Sciences Mathématiques,* 2d ser., vol. 26 (September, 1902), pp. 249–272, and also, with some modification of the more technical passages, in the *Journal des Savants* for 1902 (May), pp. 252–271. The present translation (except the postscript) is from the version in Darboux's *Bulletin,* the heavy faced figures in brackets indicating the pages of the original; the postscript appeared only in the *Journal des Savants* (p. 271).

The Grundlagen der Geometrie, by Professor David Hilbert (Leipzig, Teubner, 8vo, 92 pp.), appeared in 1899, and was reviewed for the BULLETIN by Dr. Sommer (vol. 6, 1900, pp. 287–299). A French translation by Professor L. Laugel and an English translation by Professor E. J. Townsend appeared in 1900 and 1902, and were reviewed by Dr. E. R. Hedrick in the BULLETIN, vol. 9(1902), pp. 158–165. See also a review by Mr. O. Veblen in the *Monist,* vol. 13 (January, 1903), pp. 303–309.

Reprinted from Bull. Amer. Math. Soc. **10** (October 1903), 1–23

of Euclid then cannot be demonstrated ; and this impossibility is as absolutely certain as any mathematical truth whatsoever — a fact which does not prevent the Académie des Sciences from receiving every year several new proofs, to which it naturally refuses the hospitality of the *Comptes rendus*.

Much has already been written on the non-euclidean geometries ; once they scandalized us ; now we have become accustomed to their paradoxes ; some people have gone so far as to doubt the truth of the postulate and to ask whether real space is plane, as Euclid assumed, or whether it may not present a slight *curvature*. They even supposed that experiment could give them an answer [250] to this question. Needless to add that this was a total misconception of the nature of geometry, which is not an experimental science.

But why, among all the axioms of geometry, should this postulate be the only one which could be denied without offence to logic ? Whence should it derive this privilege ? There seems to be no good reason for this, and many other conceptions are possible.

However, many contemporary geometers do not appear to think so. In recognizing the claims of the two new geometries they feel doubtless that they have gone to the extreme limit of possible concessions. It is for this reason that they have conceived what they call *general geometry*, which includes as special cases the three systems of Euclid, Lobachevsky, and Riemann, and does not include any other. And this term *general* indicates clearly that, in their minds, no other geometry is conceivable.

They will lose this illusion if they read the work of Professor Hilbert. In it they will find the barriers behind which they have wished to confine us broken down at every point.

To understand well this new attempt we must recall what has been the evolution of mathematical thought for the last hundred years, not only in geometry, but in arithmetic and in analysis. The concept of number has been made more clear and precise ; at the same time it has been generalized in various directions. The most valuable of these generalizations for the analyst is the introduction of *imaginaries* which the modern mathematician could not now dispense with ; but we have not stopped with this ; other generalizations of number, or, as we say, other categories of complex numbers, have been introduced into science.

The operations of arithmetic have in their turn been subjected to criticism, and Hamilton's quaternions have given us an example of an operation which presents an almost perfect analogy to multiplication, and may be called by the same name, which, however, is not commutative, that is, the product of two factors is not the same when the order of the factors is reversed. This was a revolution in arithmetic quite comparable to that which Lobachevsky effected in geometry.

Our conception of the infinite has been likewise modified [251]. Professor G. Cantor has taught us to distinguish gradations in infinity itself (which have, however, nothing to do with the infinitesimals of different orders invented by Leibniz for the ordinary infinitesimal calculus). The concept of the continuum, long regarded as a primitive concept, has been analyzed and reduced to its elements.

Shall I mention also the work of the Italians, who have endeavored to construct a universal logical symbolism and to reduce mathematical reasoning to purely mechanical rules?

We must recall all this if we wish to understand how it is possible that conceptions which would have staggered Lobachevsky himself, revolutionary as he was, can seem to us to-day almost natural, and can be propounded by Professor Hilbert with perfect equanimity.

THE LIST OF AXIOMS. — The first thing to do was to enumerate all the axioms of geometry. This was not so easy as one might suppose ; there are the axioms which one sees and those which one does not see, which are introduced unconsciously and without being noticed. Euclid himself, whom we suppose an impeccable logician, frequently applies axioms which he does not expressly state.

Is the list of Professor Hilbert final ? We may take it to be so, for it seems to have been drawn up with care. The distinguished professor divides the axioms into five groups :

I. Axiome der Verknüpfung (I shall translate by *projective axioms* [*axiomes projectifs*] instead of trying to find a literal translation, as for example *axioms of connection* [*axiomes de la connection*], which would not be satisfactory).

II. Axiome der Anordnung (axioms of order [*axiomes de l'ordre*]).

III. Axiom of Euclid.

IV. Axioms of congruence or metrical axioms.

V. Axiom of Archimedes.

Among the projective axioms, we shall distinguish those of the plane and those of space; the first are those derived from the familiar proposition : *through two points passes one and only one straight line;* — but I prefer to translate literally, in order to make Professor Hilbert's thought well understood.

" Let us suppose three systems of objects which we shall call *points,* [252] *straight lines,* and *planes.* Let us suppose that these points, straight lines, and planes are connected by certain relations which we shall express by the words *lying on, between,* etc.

" I. — 1.   Two different points A and B determine always a straight line *a*; in notation

$$AB = a \quad \text{or} \quad BA = a.$$

" In place of the word *determine* we shall employ as well other turns of phrase which shall be synonymous; we shall say : A lies on *a*, A is a point of *a*, *a* passes through A, *a* joins A and B, etc.

" I. — 2.   Any two points of a straight line determine this straight line; that is, if $AB = a$ and $AC = a$, and if B is different from C, we have also $BC = a$."

The following are the considerations which these statements are intended to suggest: the expressions *lying on, passing through,* etc., are not meant to call up mental pictures; they are simply synonyms of the word *determine.* The words *point, straight line,* and *plane* themselves are not intended to arouse in the mind any visual image [*représentation sensible*]. They might denote indifferently objects of any sort whatever, provided one could establish among these objects a correspondence such that to every pair of the objects called *points* there would correspond one and only one of the objects called *straight lines.* And this is why it becomes necessary to add (I, 2) that, if the line which corresponds to the pair of points A and B is the same as that which corresponds to the pair of points B and C, it is also the same as that which corresponds to the pair of points A and C.

Thus Professor Hilbert has, so to speak, sought to put the axioms into such a form that they might be applied by a person who would not understand their meaning because he had never seen either point or straight line or plane.   It should be possible, according to him, to reduce reasoning to purely mechanical rules, and it should suffice, in order to create geometry, to apply these rules slavishly to the axioms without knowing what the axioms mean. We shall thus be able to construct all geometry, I will not say precisely without understanding it at all, since we shall grasp the logical connection of the [253] propositions, but at any rate without seeing it at all. We might put the axioms into a reasoning apparatus like the logical machine* of Stanley Jevons, and see all geometry come out of it.

This is the same consideration that has inspired certain Italian scholars, such as Peano and Padoa, who have endeavored to create a *pasigraphy*, that is, a sort of universal algebra, where all the processes of reasoning are replaced by symbols or formulas.

This notion may seem artificial and puerile; and it is needless to point out how disastrous it would be in teaching and how hurtful to mental development; how deadening it would be for investigators, whose originality it would nip in the bud. But, as used by Professor Hilbert, it explains and justifies itself, if one remembers the end pursued. Is the list of axioms complete, or have we overlooked some which we apply unconsciously? This is what we want to know. For this we have one criterion, and only one. We must find out whether geometry is a logical consequence of the axioms explicitly stated, that is, whether, if we put these axioms into the reasoning machine, we can make the whole sequence of propositions come out.

If we can, we shall be sure that nothing has been overlooked. For our machine cannot work except according to the rules of logic for which it has been constructed; it ignores the vague instinct which we call *intuition*.

I shall not enlarge upon the projective axioms of space, which the author numbers I, 3, 4, 5, 6. Nothing is changed from the usual statements.

A word only on the axiom I, 7, which is thus formulated:

"On every straight line there are at least two points; on every plane there are at least three points not in a straight line; in space there are at least four points which are not in the same plane."

This statement is characteristic. Any one who had left any place for intuition, however small it might be, would not have

---

* [*Cf. Lond. Phil. Trans.*, vol. 160 (1870), pp. 497–518. *Tr.*]

dreamed of saying that on every straight line there are at least two points, or rather he would have added at once that there are an infinite number of them; for the intuition of the [254] straight line would have revealed to him both facts immediately and simultaneously.

Let us pass to the second group, that of the axioms of order. Here is the statement of the first two :

"If three points are on the same straight line, there is a certain relation among them which we express by saying that one of the points, and only one, is between the other two.   If C is between A and B, and D between A and C, then D will be also between A and B, etc."

Here again we do not bring in our intuition; we are not seeking to fathom what the word *between* may signify; every relation which satisfies the axioms might be denoted by the same word.   This is an illuminating example of the purely formal nature of mathematical definitions; but I do not dwell upon it, since I should have simply to repeat what I have said already, in speaking of the first group.

But another consideration forces itself upon us.   The axioms of order are presented as dependent on the projective axioms, and they would not have any meaning if we did not admit these latter, since we should not know what are three points on a straight line.   And nevertheless there exists a special geometry, purely qualitative, which is entirely independent of projective geometry, and does not assume the idea of the straight line, nor that of the plane, but only the ideas of curves and surfaces; this is what is called *analysis situs*.   Would it not be preferable to give to the axioms of the second group a form which would free them from this dependence and separate them completely from the first group?   It remains to be seen whether this would be possible, while preserving the purely logical character of these axioms, that is, while closing the door completely against all intuition.

The third group contains only a single axiom, which is the famous postulate of Euclid; I shall note simply that, contrary to the usual custom, it is presented before the metrical axioms.

These last form the fourth group.   We shall divide them into three subgroups.   The propositions IV, 1, 2, 3 are the metrical axioms for segments: these axioms serve to define length.   We shall agree to say that a segment taken on a [255] straight line may be congruent (equal) to a segment taken on another straight line; this is axiom IV, 1; but this convention is not wholly arbitrary; it must be so made that two segments congruent to the same third segment shall be congruent to each other (IV, 2). In the next place we define the addition of segments, by a new convention; and this convention, in turn, must be so made that when we add equal segments we find the sums equal; and this is axiom IV, 3.

The propositions IV, 4, 5 are the corresponding axioms for angles. But these are not yet sufficient; to the two subgroups of metrical axioms for segments and for angles we must add the metrical axiom for triangles (which Professor Hilbert numbers IV, 6): if two triangles have an equal angle included between equal sides, the other angles of these two triangles are equal each to each.

We recognize here one of the well known cases of equality of triangles, which we usually demonstrate by superposition, but which we must set up as a postulate if we wish to avoid making appeal to intuition. Moreover, when we made use of intuition, that is of superposition, we saw by the same process that the third sides were equal in the two triangles, and these two propositions were united, so to speak, in a single apperception; here, on the contrary, we separate them; one of them we make a postulate, but we do not set up the other as a postulate, since it can be logically deduced from the first.

Another comment: Professor Hilbert says distinctly that the segment AB is congruent to itself, but (and the same is true for angles) he should have added, should he not, that it is congruent to the inverse segment BA. This axiom (which implies the symmetry of space) is not identical with those which are explicitly stated. I do not know whether it could be logically deduced from them; I believe it could, but, given the course of reasoning of Professor Hilbert, it seems to me that this postulate is applied without being stated (page 17, line 18).

I also regret that, in this exposition of the metrical axioms, there remains no trace of an idea whose importance Helmholtz was the first to understand: I refer to the displacement of a rigid figure. It would have been possible to preserve this idea in its natural rôle, without sacrificing the logical character of the axioms. One might have said, for example: I define between figures a [256] certain relation which I call *congruence*, etc.; two figures which are congruent to the same third figure are congruent to each other; two congruent figures are identical when three points of one, not in a straight line, are identical with three corresponding points of the other, etc. The artificial introduction of this axiom IV, 6 would thus have been avoided, and the postulates would have been brought into connection with their actual psychological origin.

The fifth group contains only a single axiom, that of Archimedes.

Let A and B be any two points on a straight line D; let a be any segment; starting from the point A, and in the direction AB, construct on D a series of segments, all equal to each other and equal to a: $AA_1, A_1A_2, \cdots, A_{n-1}A_n$; then we shall always be able to take n so great that the point B will be found on one of these segments.

That is to say, if we have given any two lengths l and L, we can always find a whole number n so great that when we add the length l to itself n times, we obtain a total length greater than L.

INDEPENDENCE OF THE AXIOMS. — The list of axioms once drawn up, we must see whether it is free from contradiction. We know well that it is, since geometry exists; and Professor Hilbert also answers in the affirmative, by constructing a geometry. But this geometry, strange to say, is not quite the same as ours, his space is not our space, or at least is only a part of it. In the space of Professor Hilbert we do not have all the points which there are in our space, but only those which we can construct by ruler and compass, starting from two given points. In this space, for example, there would not exist, in general, an angle which would be the third part of a given angle.

I have no doubt that this conception would have been regarded by Euclid as more rational than ours. At any rate it is not ours. To come back to our geometry it would be necessary to add an axiom :

" If, on a straight line, there is a double infinity of points $A_1, A_2, \cdots, A_n, \cdots$; $B_1, B_2, \cdots, B_n, \cdots$, such that B_q is included between A_p and [257] B_{q-1}, and A_p between B_q and A_{p-1}, whatever the values of p and q, then there will be on this straight line at least one point C which lies between A_p and B_q, whatever the values of p and q."·

We must ask next whether the axioms are independent, that is, whether we could sacrifice one of the five groups, retaining the other four, and still attain a coherent geometry. Thus by suppressing group III (the postulate of Euclid), we obtain the non-euclidean geometry of Lobachevsky. In the same way, we can suppress group IV.

Professor Hilbert has succeeded in retaining groups I, II, III and V, along with the two subgroups of metrical axioms for segments and for angles, while rejecting the metrical axiom for triangles, that is, proposition IV, 6.

This is how he accomplishes it : consider, for simplicity, plane geometry, and let P be the plane in which we operate ; we shall retain the usual meaning for the words *point* and *straight line*, and also the usual measurement of angles ; but not so for lengths. A length shall be measured *by definition* by its projection on a plane Q different from P, this projection itself being measured in the usual way. It is clear that all the axioms will hold, except the metrical axioms. The metrical axioms for angles will also hold, since we change nothing concerning the measurement of angles ; those for segments will also hold, since each segment is measured by another segment which is its projection on the plane Q, and this latter segment is measured in the usual way. On the other hand, the theorems on the equality of triangles, such as the axiom IV, 6, are no longer true. This solution satisfies me only half-way ; angles have been defined independently of lengths, without trying to bring the two definitions into agreement (or rather, by bringing them purposely into disagreement). To return to classic geometry it would be sufficient to change *one* of the two definitions. I should prefer to have had the lengths so defined as to make it impossible to find a definition of angles satisfying the metrical axioms for angles and for triangles. This would moreover not be difficult [258].

It would have been easy for Professor Hilbert to create a geometry in which the axioms of order would be abandoned while all the others would be retained. Or rather this geometry exists already, or rather there exist two of them. There is that of Riemann, for which, it is true, the postulate of Euclid (group III) is also abandoned, since the sum of the angles of a triangle is greater than two right angles. To make my thought clear I shall limit myself to considering a geometry of two dimensions. The geometry of Riemann in two dimensions is nothing else than spherical geometry, with one condition, namely, that we shall not regard as distinct two diametrically opposite points on the sphere. The elements of this geometry will then be the different diameters of this sphere. Now, if we consider three diameters of the same sphere, lying in the same diametral plane, we have no reason for saying that one of them is *between* the other two. The word *between* has no longer any meaning, and the axioms of order drop out of themselves.

If we wish now a geometry in which the axioms of order shall not hold, while the axiom of Euclid is retained with the others, we have only to take as elements the *imaginary* points and straight lines in ordinary space. It is clear that the imaginary points of space are not given us as *arranged* in a definite order. But more than that: we may ask whether they are capable of being so arranged; this would undoubtedly be possible, as G. Cantor has shown (subject to the condition, be it understood, of not always arranging in close proximity points which we regard as infinitely near, and of destroying thereby the continuity of space). We might, I say, arrange them, but this could not be done in such a way that the arrangement would not be altered by the various operations of geometry (projection, translation, rotation, etc.). The axioms of order, then, are not applicable to this geometry.

THE NON-ARCHIMEDEAN GEOMETRY. — But the most original conception of Professor Hilbert is that of non-archimedean geometry, in which all the axioms remain true except that of Archimedes. For this it was necessary, in the first place, to construct a *system* [259] *of non-archimedean numbers*, that is, a system of elements among which we may define the relations of equality and inequality and to which we may apply operations analogous to arithmetical addition and multiplication — and this in such a way as to satisfy the following conditions :

1° The arithmetical rules for addition and multiplication (the commutative, associative, distributive laws, etc.: *Arithmetische Axiome der Verknüpfung*) hold without change.

2° The rules for the establishment and transformation of inequalities (*Arithmetische Axiome der Anordnung*) likewise hold.

3° The axiom of Archimedes is not true.

We may reach this result by choosing for elements series of the following form :

$$A_0 t^m + A_1 t^{m-1} + A_2 t^{m-2} + \cdots,$$

where m is a positive or negative integer and where the coefficients A are real, and by agreeing to apply to these series the ordinary rules of addition and multiplication. We must then define the conditions of inequality of these series, so as to *arrange* our elements in a definite order. We shall accomplish this by the following convention : we shall give to our series the sign of A_0 and we shall say that one series is less than another when, subtracted from the other, it leaves a positive remainder.

It is clear that with this convention the rules of the calculus of inequalities hold; but the axiom of Archimedes is no longer true; for, if we take the two elements 1 and t, the first added to itself as many times as we please remains always less than the second. We shall have always $t > n$, whatever the whole number n, since the difference $t - n$ will always be positive; for the coefficient of the first term t, which, by definition, gives its sign, remains always equal to 1.

Our ordinary numbers come in as particular cases among these *non-archimedean numbers.* The new numbers are interpolated, so to speak, in the series of our ordinary numbers, in such a way that we may have, for example, an infinity of the new numbers less than a given ordinary number A and greater than all the ordinary numbers less than A [260].

This premised, imagine a space of three dimensions in which the coordinates of a point would be measured not by ordinary numbers but by non-archimedean numbers, while the usual equations of the straight line and the plane would hold, as well as the analytic expressions for angles and lengths. It is clear that in this space all the axioms would remain true except that of Archimedes.

On every straight line new points would be interpolated between our ordinary points. If, for example, D_0 is an ordinary straight line, and D_1 the corresponding non-archimedean straight line; if P is any ordinary point of D_0, and if this point divides D_0 into two half-rays S and S' (I add, for precision, that I consider P as not belonging to either S or S'); then there will be on D_1 an infinity of new points as well between P and S as between P and S'. There will be also on D_1 an infinity of new points which will lie to the right of all the ordinary points of D_0. In short, our ordinary space is only a part of the non-archimedean space.

At the first blush the mind revolts against conceptions like this.   This is because, through an old habit, it is looking for a visual image.   It must free itself from this prejudice if it would arrive at comprehension, and this is even more necessary here than in the case of non-euclidean geometry.   Professor Hilbert has only one object in view : to construct a system of elements capable of certain logical relations ; and it is sufficient for him to show that these relations do not involve any self-contradiction.

We may remark in passing that the non-euclidean geometry respects, so to speak, our qualitative conception of the geometrical continuum, while entirely overturning our ideas about the measurement of this continuum.   The non-archimedean geometry destroys this concept, by dissecting the continuum for the introduction of new elements.

Whatever they may be, Professor Hilbert follows out the consequences of his premises and tries to see how one could remake geometry without using the axiom of Archimedes. There is no difficulty in the chapters which the school-boys call the *first* and *second* Books.   This axiom does not occur at any point in those Books.

The third Book treats of proportions and of similarity.   The plan which Professor Hilbert follows for the [261] reconstruction of this book without recourse to the axiom of Archimedes is, in substance, as follows.   He takes the usual construction of the fourth proportional as the definition of proportion ; but such a definition needs to be justified ; he needs to show in the first place that the result is the same whatever may be the auxiliary lines employed in the construction, and next that the ordinary rules of operation apply to the proportions thus defined.   This justification Professor Hilbert gives us in a satisfactory manner.

The fourth Book treats of the measurement of plane areas. If this measurement can be easily established without the aid of the principle of Archimedes, it is because two equivalent polygons can either be decomposed into triangles in such a way that the component triangles of the one and those of the other are equal each to each (or, in other words, can be converted one into the other after the manner of the Chinese puzzle *), or else can be regarded as the difference of polygons capable of this mode of decomposition (this is really the same process, admit-

---

* [By cutting up and putting together again.   *Tr.*]

ting not only positive triangles but also negative triangles). But we must observe that an analogous state of affairs does not seem to exist in the case of two equivalent polyhedra, so that it becomes a question whether we can determine the volume of the pyramid, for example, without an appeal more or less disguised to the infinitesimal calculus. It is then not certain whether we could dispense with the axiom of Archimedes as easily in the measurement of volumes as in that of plane areas. Moreover Professor Hilbert has not attempted it.

One question remains to be treated in any case ; a polygon being given, is it possible to cut it up into triangles and remove one of the pieces in such a way that the remaining polygon may be equivalent to the given polygon, that is to say, in such a way that by transforming this remaining polygon by the process of the Chinese puzzle we could come back to the original polygon? Ordinarily we are satisfied with saying that this is impossible because the whole is greater than the part. This is to call in a new axiom, and, however obvious it may seem to us, the logician would be better satisfied if we could avoid it. Professor Schur has discovered a proof, it is true, but it depends on the axiom of Archimedes ; Professor Hilbert wished to reach the result without using this axiom. This is the device by which he [262] does it : he adopts as the *definition* of the *area* of the triangle half the product of its base by its altitude, and he justifies this definition by showing that two triangles which are equivalent (from the point of view of the Chinese puzzle) have the same *area* (in the sense of the new definition) and that the *area* of a triangle which can be decomposed into several others is the sum of the *areas* of the component triangles. This justification once out of the way, all the rest follows without difficulty. It is always the same process. To avoid constant appeals to intuition, which would provide us constantly with new axioms, we change these axioms into definitions, and afterwards justify these definitions by showing that they are free from contradictions.

THE NON-ARGUESIAN GEOMETRY. — The fundamental theorem of projective geometry is the theorem of Desargues. Two triangles are called *homologous* when the straight lines which join the corresponding vertices intersect in the same point. Desargues has shown that the points of intersection of the corresponding sides of two homologous triangles are on the same straight line ; the converse is also true.

The theorem of Desargues can be established in two ways :

1° By using the projective axioms of the plane and the metrical axioms of the plane.

2° By using the projective axioms of the plane and those of space.

The theorem might then be discovered by a two-dimensional animal, to whom a third dimension would seem as inconceivable as a fourth does to us ; such an animal would then be ignorant of the projective axioms of space ; but he would have seen movement, in the plane which he inhabits, of rigid figures analogous to our rigid bodies, and would consequently be acquainted with the metrical axioms. The theorem could be discovered also by a three-dimensional animal who was acquainted with the projective axioms of space, but who, never having seen rigid bodies move, would be ignorant of the metrical axioms.

But could we establish the theorem of Desargues without using either the projective axioms of space or the metrical axioms, [263] but only the projective axioms of the plane ? We thought not, but we were not sure of it. Professor Hilbert has decided the question by constructing a *non-arguesian geometry*, which is, of course, a plane geometry. Consider an ellipse E. Outside of this ellipse the word *straight line* preserves its ordinary meaning : in the interior the word *straight line* takes a different meaning and denotes an arc of a circle which, when produced, would pass through a fixed point P outside the ellipse. A straight line which crosses the ellipse E is then composed of two rectilinear parts, in the ordinary sense of the word, connected in the interior of the ellipse by an arc of a circle ; like a ray of light which would be deflected from its rectilinear path by passing through a refracting body.

The projective axioms of the plane will still be true if we take the point P sufficiently far removed from the ellipse E.

Now place two homologous triangles outside the ellipse E, and in such a way that their sides do not meet E ; the three straight lines which join the corresponding vertices two and two, *if we take them in the ordinary sense of the word*, will meet in the same point Q, according to the theorem of Desargues ; suppose that this point Q is in the interior of E. *If we take the word* straight line *in the new sense*, the three straight lines which join the corresponding vertices will be deflected on entering the interior of the ellipse. They will then no longer pass through Q, they will be no longer concurrent. The theorem of Desargues is no longer true in our new geometry; this is a non-arguesian geometry.

THE NON-PASCALIAN GEOMETRY. — Professor Hilbert does not stop here, but introduces still another new conception. In order to understand it, we must first return a moment into the domain of arithmetic. We have noticed above the extension of the concept of number, by the introduction of the *non-archimedean numbers*. We want a classification of these new numbers, to obtain which we shall begin by classifying the axioms of arithmetic in four groups, which are:

1° The associative and commutative laws of addition, the associative law of multiplication, the two [264] distributive laws of multiplication; or, in short, all the rules of addition and of multiplication, except the commutative law of multiplication;

2° The axioms of order; that is, the rules of the calculus of inequalities;

3° The commutative law of multiplication, according to which we can invert the order of the factors without changing the product;

4° The axiom of Archimedes.

Numbers which admit the axioms of the first two groups shall be called *arguesian;* they may be *pascalian* or *non-pascalian*, according as they satisfy or do not satisfy the axiom of the third group; they will be *archimedean* or *non-archimedean*, according as they satisfy or not the axiom of the fourth group. We shall soon see the reason for these names.

Ordinary numbers are at the same time arguesian, pascalian and archimedean. It can be shown that the commutative law follows from the axioms of the first two groups and the axiom of Archimedes; there are therefore no numbers which are arguesian, archimedean and not pascalian.

On the other hand, we have cited above an example of numbers which were arguesian, pascalian and not archimedean; I shall call these the *numbers of the system* T, and I recall that to each of these numbers there corresponds a series of the form

$$A_0 t^m + A_1 t^{m-1} + \cdots,$$

where the A's are ordinary real numbers.

It is easy to construct, by an analogous process, a system of arguesian numbers which are non-pascalian and non-archimedean. The elements of this system will be series of the form

$$S = T_0 s^n + T_1 s^{n-1} + \cdots ,$$

where s is a symbol analogous to t, n a positive or negative integer, and T_0, T_1, $\cdots$ numbers of the system T; if then we replaced the coefficients T_0, T_1, $\cdots$ by the corresponding series in t we should have a series depending on both t and s. We shall [265] add these series S according to the ordinary rules; also for the multiplication of these series we shall admit the distributive and associative laws; but we shall suppose that the commutative law is not true, that on the contrary $st = - ts$.

It remains to *arrange* the series in a definite order, so as to satisfy the axioms of order. For this, we shall attribute to the series S the sign of the first coefficient T_0; we shall say that one series is less than another when, subtracted from the first, it leaves a positive remainder. It is always the same scheme: t is regarded as very great in comparison with any ordinary real number, and s is regarded as very great in comparison with any number of the system T.

The commutative law not being true, these are clearly non-pascalian numbers.

Before going farther, I recall that Hamilton introduced long ago a system of complex numbers in which multiplication is not commutative; these are the *quaternions*, of which the English make such frequent use in mathematical physics. But, in the case of quaternions, the axioms of order are not true; the originality of Professor Hilbert's conception lies in this, that his new numbers satisfy the axioms of order without satisfying the commutative law.

To return to geometry. Admit the axioms of the first three groups, that is, the projective axioms of the plane and of space, the axioms of order and the postulate of Euclid: the theorem of Desargues will follow from them, since it is a consequence of the projective axioms of space.

We wish to construct our geometry *without making use of the metrical axioms*; the word *length* has then for us no meaning; we have no right to use the compass; on the other hand, we may use the ruler, since we admit that we can draw a straight line through two points, by virtue of one of the projective axioms;

also, we know how to draw through a given point a parallel to a given straight line, since we admit the postulate of Euclid. Let us see what we can do with these resources.

We can define the *homothetic relation* * [*homothétie*] of two figures; two triangles shall be called homothetic when their sides are [266] parallel two and two, and we conclude from this (by the theorem of Desargues, which we admit) that the straight lines which join the corresponding vertices are concurrent. We shall then make use of the homothetic relation to define proportion. We can also define equality to a certain extent.

Two opposite sides of a parallelogram shall be equal *by definition;* we can thus decide whether two segments are equal to each other, provided they are parallel.

Thanks to these conventions, we are now in a position to compare the lengths of two segments; *provided*, however, *that these segments are* PARALLEL. The comparison of two lengths which have different directions has no meaning; there would be required, so to speak, a different unit of length for each direction. Needless to add that the word *angle* has no meaning.

Lengths will thus be expressed by numbers; but these will not necessarily be ordinary numbers. All that we can say is this, that, if the theorem of Desargues is true, as we admit, these numbers will belong to a system satisfying the arithmetic axioms of the first two groups, that is, to an *arguesian system.* Conversely, being given any system S of arguesian numbers, we can construct a geometry in which the lengths of segments of a straight line can be exactly expressed by these numbers.

Here is the way in which this can be done: a point of this new space shall be *defined* by three numbers x, y, z of the system S which we shall call the *coordinates* of this point. If to the three coordinates of the various points of a figure we add three constants (which are, of course, arguesian numbers of the system S), we obtain another figure, derived from the first in such a way that to any segment of one of the figures there corresponds an equal and parallel segment in the other (in the sense given above to this word). This transformation is then a translation, so that these three constants might define a translation. If now we multiply the three coordinates of all the points of a given figure by the same constant, we shall obtain a second figure which will be homothetic to the first [267].

---

* [Two figures are *homothetic* when they are similar and similarly placed. *Tr.*]

The equation of a plane will be the well known linear equation of ordinary analytic geometry; but, since in the system S multiplication will not in general be commutative, it is important to make a distinction and to say that in each of the terms of this linear equation that factor shall be the coordinate which plays the rôle of multiplicand, and that the constant coefficient which plays the rôle of multiplier.

Thus, to each system of arguesian numbers there will correspond a new geometry satisfying the projective axioms, the axioms of order, the theorem of Desargues, and the postulate of Euclid. What is now the geometric meaning of the arithmetic axiom of the third group, that is, of the commutative law of multiplication? *Translated into geometric language, this law is the theorem of Pascal;* I refer to the theorem on the hexagon inscribed in a conic, supposing that this conic reduces to two straight lines.

Thus the theorem of Pascal will be true or false according as the system S is pascalian or non-pascalian; and, since there are non-pascalian systems, *there are also non-pascalian geometries.*

The theorem of Pascal can be proved by starting with the metrical axioms; it will then be true, if we admit that figures can be transformed not only by the homothetical transformation and translation, as we have just been doing, but also by rotation.

The theorem of Pascal can also be deduced from the axiom of Archimedes, since we have just seen that every system of numbers which is arguesian and archimedean is at the same time pascalian; *every non-pascalian geometry is then at the same time non-archimedean.*

The Streckenübertrager. — Let us mention one more idea of Professor Hilbert. He studies the constructions which can be made, not with the aid of the ruler and compass, but by means of the ruler and a special instrument which he calls *Streckenübertrager*, and which would enable us to lay off on a straight line a segment equal to another segment taken on another straight line. The *Streckenübertrager* is not equivalent to the compasses; this latter instrument would enable us to construct the point of intersection of two circles or of a [268] circle and any straight line; the *Streckenübertrager* would give us only the intersection of a circle with a straight line *passing through the center of the circle.*   Professor Hilbert inquires then what constructions will be possible with these two instruments, and he reaches a quite remarkable conclusion.

The constructions which can be made by the ruler and compass can also be made by the ruler and the *Streckenübertrager*, *if these constructions are such that their result is always real.*   It is easy enough to see that this condition is necessary ; for a circle is always cut in two real points by a straight line drawn through its centre.   But it was difficult to foresee that this condition would be also sufficient.

Various Geometries. — I should like, before closing, to see what places are taken in Professor Hilbert's classification by the various geometries which have been proposed up to the present time.   In the first place, the geometries of Riemann ; I do not mean *the* geometry of Riemann which has been mentioned above and which is contrasted with that of Lobachevsky ; I mean the geometries connected with space of variable curvature considered by Riemann in his celebrated *Habilitationsschrift.*

In this conception, any curve has a length assigned to it, by definition, and it is on this definition that everything depends. The rôle of straight lines is played by the geodesics, that is, by the lines of minimum length drawn from one point to another. The projective axioms are no longer true, and there is no reason why, for example, two points could not be joined by more than one geodesic.   The postulate of Euclid clearly can no longer have any meaning.   The axiom of Archimedes remains true, as well as the axioms of order, *mutatis mutandis ;* Riemann does not consider, indeed, any but the ordinary system of numbers.   As to the metrical axioms, it is easily seen that those for segments and those for angles remain true, while the metrical axiom for triangles (IV, 6) is evidently false.

And here we meet the objection which has been most often made to Riemann.

" You speak of length," they say to him ; " now length assumes measurement, and to measure we must be able to carry about a measuring [269] instrument which must remain invariant ; moreover, you recognize this yourself.   Space then must be everywhere equal to itself, it must be homogeneous in order that congruence may be possible.   Now your space is not homogeneous, since its curvature is variable ; in such a space there can be no such thing as measurement or length."

Riemann would have had no trouble in replying. Consider, for simplicity, a geometry of two dimensions; we shall be able then to picture to ourselves Riemann's space as a surface in ordinary space. We might measure lengths on this surface by means of a thread, and nevertheless a figure could not be moved about in this surface in such a way that the lengths of all its elements remain invariant. For the surface is not, in general, applicable on itself.

This is what Professor Hilbert would express by saying that the metrical axioms for segments are true, while that for triangles is not. The first find concrete expression, so to speak, in our thread; the axiom for triangles would assume a displacement of a figure all of whose elements would have a constant length.

What will be the place of another geometry which I have proposed on a former occasion * and which belongs, so to speak, to the same family as that of Lobachevsky and that of Riemann? I have shown that we can imagine three geometries in two dimensions, which correspond respectively to three kinds of surfaces of the second degree : the ellipsoid, the hyperboloid of two sheets, and the hyperboloid of one sheet ; the first is that of Riemann, the second is that of Lobachevsky, and the third is the new geometry. We should find in the same way four geometries in three dimensions.

Where would this new geometry stand in the classification of Professor Hilbert? It is easy to discover. As in the case of the geometry of Riemann, all the axioms hold, save those of order and that of Euclid; but, while in the geometry of Riemann the axioms are false on all the straight lines, in the new geometry, on the contrary, the straight lines separate themselves into two classes, those on which the axioms of order are true, and those on which they are false [270].

Conclusions. — But the most important thing is to arrive at a clear understanding of the place which the new conceptions of Professor Hilbert occupy in the history of our ideas on the philosophy of mathematics.

After a first period of naïve confidence in which we cherished the hope of demonstrating everything, came Lobachevsky, the inventor of the non-euclidean geometries.

---

* [See *Bull. de la Société mathématique de France*, vol. 15 (1887), pp. 203–216. Other articles by Poincaré on the foundations of geometry have appeared in the *Revue de Métaphysique*, vol. 7, and the *Monist*, vol. 9. *Tr.*]

But the true meaning of this discovery was not fathomed all at once; Helmholtz showed in the first place that the propositions of euclidean geometry were no other than the laws of motion of rigid bodies, while the propositions of the other geometries were the laws which might govern other bodies analogous to the rigid bodies — bodies which doubtless do not exist, but whose existence might be conceived without leading to the least contradiction, bodies which we might fabricate if we wished. These laws could not, however, be regarded as experimental, since the solids of nature follow them only roughly, and since, besides, the fictitious bodies of non-euclidean geometry do not exist, and cannot be accessible to experiment. Helmholtz, moreover, never explained himself altogether clearly on this point.

Lie pushed the analysis much farther. He inquired in what way the various possible movements of any system, or more generally the various possible transformations of a figure, can be combined. If we consider a certain number of transformations, and suppose that they are combined in all possible ways, the totality of all these combinations will form what he calls *a group*. To each group corresponds a geometry, and ours, which corresponds to the group of displacements of a rigid body, is only a very special case. But all the groups which one can imagine will possess certain common properties, and it is precisely these common properties which limit the caprice of the inventors of geometries; it is they, indeed, which Lie studied all his life.

He was, however, not entirely satisfied with his work. He had, he said, always regarded space as a *Zahlenmannigfaltigkeit*. He had confined himself to the study of continuous groups [271] properly so called, to which the rules of the ordinary infinitesimal analysis apply. Was he not thus artificially restricted? Had he not thus neglected one of the indispensable axioms of geometry (referring to the axiom of Archimedes)? I do not know whether any trace of this thought would be found in his printed works, but in his correspondence, or in his conversation, he constantly expressed this same concern.

This is precisely the gap which Professor Hilbert has filled up; the geometries of Lie remained all subject to the forms of analysis and of arithmetic, which seemed unassailable. Professor Hilbert has broken through these forms, or, if you prefer, he has enlarged them. His spaces are no longer *Zahlenmannigfaltigkeiten*.

The objects which he calls points, straight lines, or planes become thus purely logical entities which it is impossible to represent to ourselves. We should not know how to picture them as sensory images, these points which are nothing but systems of three series. It matters little to him ; it is sufficient for him that they are *individuals* and that he has positive rules for distinguishing these individuals one from another, for establishing arbitrarily between them relations of equality or of inequality, and for transforming them.

One other comment : the groups of transformations in Lie's sense appear to play only a secondary part. At least this is how it seems when we read the actual text of Professor Hilbert. But, if we should consider it more closely, we should see that each of his geometries is still the study of a group. His non-archimedean geometry is the study of a group which contains all the transformations of the euclidean group, corresponding to the various displacements of a rigid body, but which contains also other transformations capable of being combined with the first according to simple laws.

Lobachevsky and Riemann rejected the postulate of Euclid, but they preserved the metrical axioms ; in the majority of his geometries, Professor Hilbert does the opposite. This amounts to placing in the first rank a group comprising the transformations of space by the homothetic transformation and by translation ; and at the foundation of his non-pascalian geometry we meet an analogous group, comprising not only the homothetic transformation and the translations of ordinary space, but other analogous transformations which combine with the first according to simple laws [272].

Professor Hilbert seems rather to slur over these inter-relations ; I do not know why. The logical point of view alone appears to interest him. Being given a sequence of propositions, he finds that all follow logically from the first. With the foundation of this first proposition, with its psychological origin, he does not concern himself. And even if we have, for example, three propositions A, B, C, and if it is logically possible, by starting with any one among them, to deduce the other two from it, it will be immaterial to him whether we regard A as an axiom, and derive B and C from it, or whether, on the contrary, we regard C as an axiom, and derive A and B from it. The axioms are postulated ; we do not know where they come from ; it is then as easy to postulate A as C.

His work is then incomplete; but this is not a criticism which I make against him. Incomplete one must indeed resign one's self to be. It is enough that he has made the philosophy of mathematics take a long step in advance, comparable to those which were due to Lobachevsky, to Riemann, to Helmholtz, and to Lie.

———

Since * the printing of the preceding lines, Professor Hilbert has published a new note on the same subject ("Ueber die Grundlagen der Geometrie," *Nachrichten der K. Gesellschaft der Wissenschaften zu Göttingen*, 1902, Heft 3). He seems to have made here an attempt to fill in the gaps which I have noticed above. Although this note is very concise, one sees clearly two thoughts running through it. In the first place he seeks to present the axioms of order emancipated from all dependence on projective geometry; he uses for this a theorem of Professor Jordan. Next, he reconnects the fundamental principles of geometry with the notion of a group. He comes nearer then to the point of view of Lie, but he makes an advance on the work of his predecessor, since he frees the theory of groups from all appeal to the principles of the differential calculus.

H. POINCARÉ.

BULLETIN (New Series) OF THE
AMERICAN MATHEMATICAL SOCIETY
Volume 37, Number 4, Pages 407–436
S 0273-0979(00)00881-8
Article electronically published on June 26, 2000

MATHEMATICAL PROBLEMS

DAVID HILBERT

Lecture delivered before the International Congress of Mathematicians at Paris in 1900.

Who of us would not be glad to lift the veil behind which the future lies hidden; to cast a glance at the next advances of our science and at the secrets of its development during future centuries? What particular goals will there be toward which the leading mathematical spirits of coming generations will strive? What new methods and new facts in the wide and rich field of mathematical thought will the new centuries disclose?

History teaches the continuity of the development of science. We know that every age has its own problems, which the following age either solves or casts aside as profitless and replaces by new ones. If we would obtain an idea of the probable development of mathematical knowledge in the immediate future, we must let the unsettled questions pass before our minds and look over the problems which the science of to-day sets and whose solution we expect from the future. To such a review of problems the present day, lying at the meeting of the centuries, seems to me well adapted. For the close of a great epoch not only invites us to look back into the past but also directs our thoughts to the unknown future.

The deep significance of certain problems for the advance of mathematical science in general and the important rôle which they play in the work of the individual investigator are not to be denied. As long as a branch of science offers an abundance of problems, so long is it alive; a lack of problems foreshadows extinction or the cessation of independent development. Just as every human undertaking pursues certain objects, so also mathematical research requires its problems. It is by the solution of problems that the investigator tests the temper of his steel; he finds new methods and new outlooks, and gains a wider and freer horizon.

It is difficult and often impossible to judge the value of a problem correctly in advance; for the final award depends upon the grain which science obtains from the problem. Nevertheless we can ask whether there are general criteria which mark a good mathematical problem. An old French mathematician said: "A mathematical theory is not to be considered complete until you have made it so clear that you can explain it to the first man whom you meet on the street." This clearness and ease of comprehension, here insisted on for a mathematical theory, I should still more demand for a mathematical problem if it is to be perfect; for what is clear and easily comprehended attracts, the complicated repels us.

Moreover a mathematical problem should be difficult in order to entice us, yet not completely inaccessible, lest it mock at our efforts. It should be to us a guide

---

Reprinted from Bull. Amer. Math. Soc. **8** (July 1902), 437–479.

Originally published as *Mathematische Probleme. Vortrag, gehalten auf dem internationalen Mathematike-Congress zu Paris 1900*, Gött. Nachr. 1900, 253-297, Vandenhoeck & Ruprecht, Göttingen. Translated for the *Bulletin*, with the author's permission, by Dr. Mary Winston Newson, 1902.

post on the mazy paths to hidden truths, and ultimately a reminder of our pleasure in the successful solution.

The mathematicians of past centuries were accustomed to devote themselves to the solution of difficult particular problems with passionate zeal. They knew the value of difficult problems. I remind you only of the "problem of the line of quickest descent," proposed by John Bernoulli. Experience teaches, explains Bernoulli in the public announcement of this problem, that lofty minds are led to strive for the advance of science by nothing more than by laying before them difficult and at the same time useful problems, and he therefore hopes to earn the thanks of the mathematical world by following the example of men like Mersenne, Pascal, Fermat, Viviani and others and laying before the distinguished analysts of his time a problem by which, as a touchstone, they may test the value of their methods and measure their strength. The calculus of variations owes its origin to this problem of Bernoulli and to similar problems.

Fermat had asserted, as is well known, that the diophantine equation

$$x^n + y^n = z^n$$

(x, y and z integers) is unsolvable—except in certain self-evident cases. The attempt to prove this impossibility offers a striking example of the inspiring effect which such a very special and apparently unimportant problem may have upon science. For Kummer, incited by Fermat's problem, was led to the introduction of ideal numbers and to the discovery of the law of the unique decomposition of the numbers of a circular field into ideal prime factors—a law which to-day in its generalization to any algebraic field by Dedekind and Kronecker, stands at the center of the modern theory of numbers and whose significance extends far beyond the boundaries of number theory into the realm of algebra and the theory of functions.

To speak of a very different region of research, I remind you of the problem of three bodies. The fruitful methods and the far-reaching principles which Poincaré has brought into celestial mechanics and which are to-day recognized and applied in practical astronomy are due to the circumstance that he undertook to treat anew that difficult problem and to approach nearer a solution.

The two last mentioned problems—that of Fermat and the problem of the three bodies—seem to us almost like opposite poles—the former a free invention of pure reason, belonging to the region of abstract number theory, the latter forced upon us by astronomy and necessary to an understanding of the simplest fundamental phenomena of nature.

But it often happens also that the same special problem finds application in the most unlike branches of mathematical knowledge. So, for example, the problem of the shortest line plays a chief and historically important part in the foundations of geometry, in the theory of curved lines and surfaces, in mechanics and in the calculus of variations. And how convincingly has F. Klein, in his work on the icosahedron, pictured the significance which attaches to the problem of the regular polyhedra in elementary geometry, in group theory, in the theory of equations and in that of linear differential equations.

In order to throw light on the importance of certain problems, I may also refer to Weierstrass, who spoke of it as his happy fortune that he found at the outset of his scientific career a problem so important as Jacobi's problem of inversion on which to work.

Having now recalled to mind the general importance of problems in mathematics, let us turn to the question from what sources this science derives its problems. Surely the first and oldest problems in every branch of mathematics spring from experience and are suggested by the world of external phenomena. Even the rules of calculation with integers must have been discovered in this fashion in a lower stage of human civilization, just as the child of to-day learns the application of these laws by empirical methods. The same is true of the first problems of geometry, the problems bequeathed us by antiquity, such as the duplication of the cube, the squaring of the circle; also the oldest problems in the theory of the solution of numerical equations, in the theory of curves and the differential and integral calculus, in the calculus of variations, the theory of Fourier series and the theory of potential—to say noting of the further abundance of problems properly belonging to mechanics, astronomy and physics.

But, in the further development of a branch of mathematics, the human mind, encouraged by the success of its solutions, becomes conscious of its independence. It evolves from itself alone, often without appreciable influence from without, by means of logical combination, generalization, specialization, by separating and collecting ideas in fortunate ways, new and fruitful problems, and appears then itself as the real questioner. Thus arose the problem of prime numbers and the other problems of number theory, Galois's theory of equations, the theory of algebraic invariants, the theory of abelian and automorphic functions; indeed almost all the nicer questions of modern arithmetic and function theory arise in this way.

In the meantime, while the creative power of pure reason is at work, the outer world again comes into play, forces upon us new questions from actual experience, opens up new branches of mathematics, and while we seek to conquer these new fields of knowledge for the realm of pure thought, we often find the answers to old unsolved problems and thus at the same time advance most successfully the old theories. And it seems to me that the numerous and surprising analogies and that apparently prearranged harmony which the mathematician so often perceives in the questions, methods and ideas of the various branches of his science, have their origin in this ever-recurring interplay between thought and experience.

It remains to discuss briefly what general requirements may be justly laid down for the solution of a mathematical problem. I should say first of all, this: that it shall be possible to establish the correctness of the solution by means of a finite number of steps based upon a finite number of hypotheses which are implied in the statement of the problem and which must always be exactly formulated. This requirement of logical deduction by means of a finite number of processes is simply the requirement of rigor in reasoning. Indeed the requirement of rigor, which has become proverbial in mathematics, corresponds to a universal philosophical necessity of our understanding; and, on the other hand, only by satisfying this requirement do the thought content and the suggestiveness of the problem attain their full effect. A new problem, especially when it comes from the world of outer experience, is like a young twig, which thrives and bears fruit only when it is grafted carefully and in accordance with strict horticultural rules upon the old stem, the established achievements of our mathematical science.

Besides it is an error to believe that rigor in the proof is the enemy of simplicity. On the contrary we find it confirmed by numerous examples that the rigorous method is at the same time the simpler and the more easily comprehended. The very effort for rigor forces us to find out simpler methods of proof. It also frequently leads the way to methods which are more capable of development than the old methods of less rigor. Thus the theory of algebraic curves experienced a considerable simplification and attained greater unity by means of the more rigorous function-theoretical methods and the consistent introduction of transcendental devices. Further, the proof that the power series permits the application of the four elementary arithmetical operations a well as the term by term differentiation and integration, and the recognition of the utility of the power series depending upon this proof contributed materially to the simplification of all analysis, particularly of the theory of elimination and the theory of differential equations, and also of the existence proofs demanded in those theories. But the most striking example for my statement is the calculus of variations. The treatment of the first and second variations of definite integrals required in part extremely complicated calculations, and the processes applied by the old mathematicians had not the needful rigor. Weierstrass showed us the way to a new and sure foundation of the calculus of variations. By the examples of the simple and double integral I will show briefly, at the close of my lecture, how this way leads at once to a surprising simplification of the calculus of variations. For in the demonstration of the necessary and sufficient criteria for the occurrence of a maximum and minimum, the calculation of the second variation and in part, indeed, the wearisome reasoning connected with the first variation may be completely dispensed with—to say nothing of the advance which is involved in the removal of the restriction to variations for which the differential coefficients of the function vary but slightly.

While insisting on rigor in the proof as a requirement for a perfect solution of a problem, I should like, on the other hand, to oppose the opinion that only the concepts of analysis, or even those of arithmetic alone, are susceptible of a fully rigorous treatment. This opinion, occasionally advocated by eminent men, I consider entirely erroneous. Such a one-sided interpretation of the requirement of rigor would soon lead to the ignoring of all concepts arising from geometry, mechanics and physics, to a stoppage of the flow of new material from the outside world, and finally, indeed, as a last consequence, to the rejection of the ideas of the continuum and of the irrational number. But what an important nerve, vital to mathematical science, would be cut by the extirpation of geometry and mathematical physics! On the contrary I think that wherever, from the side of the theory of knowledge or in geometry, or from the theories of natural or physical science, mathematical ideas come up, the problem arises for mathematical science to investigate the principles underlying these ideas and so to establish them upon a simple and complete system of axioms, that the exactness of the new ideas and their applicability to deduction shall be in no respect inferior to those of the old arithmetical concepts.

To new concepts correspond, necessarily, new signs. These we choose in such a way that they remind us of the phenomena which were the occasion for the formation of the new concepts. So the geometrical figures are signs or mnemonic symbols of space intuition and are used as such by all mathematicians. Who does not always use along with the double inequality $a > b > c$ the picture of three points following one another on a straight line as the geometrical picture of the idea "between"? Who does not make use of drawings of segments and rectangles enclosed in one another, when it is required to prove with perfect rigor a difficult theorem on the continuity of functions or the existence of points of condensation? Who could dispense with the figure of the triangle, the circle with its center, or with the cross of three perpendicular axes? Or who would give up the representation of the vector field, or the picture of a family of curves or surfaces with its envelope which plays so important a part in differential geometry, in the theory of differential equations, in the foundation of the calculus of variations and in other purely mathematical sciences?

The arithmetical symbols are written diagrams and the geometrical figures are graphic formulas; and no mathematician could spare these graphic formulas, any more than in calculation the insertion and removal of parentheses or the use of other analytical signs.

The use of geometrical signs as a means of strict proof presupposes the exact knowledge and complete mastery of the axioms which underlie those figures; and in order that these geometrical figures may be incorporated in the general treasure of mathematical signs, there is necessary a rigorous axiomatic investigation of their conceptual content. Just as in adding two numbers, one must place the digits under each other in the right order, so that only the rules of calculation, *i.e.*, the axioms of arithmetic, determine the correct use of the digits, so the use of geometrical signs is determined by the axioms of geometrical concepts and their combinations.

The agreement between geometrical and arithmetical thought is shown also in that we do not habitually follow the chain of reasoning back to the axioms in arithmetical, any more than in geometrical discussions. On the contrary we apply, especially in first attacking a problem, a rapid, unconscious, not absolutely sure combination, trusting to a certain arithmetical feeling for the behavior of the arithmetical symbols, which we could dispense with as little in arithmetic as with the geometrical imagination in geometry. As an example of an arithmetical theory operating rigorously with geometrical ideas and signs, I may mention Minkowski's work, Die Geometrie der Zahlen.[1]

Some remarks upon the difficulties which mathematical problems may offer, and the means of surmounting them, may be in place here.

If we do not succeed in solving a mathematical problem, the reason frequently consists in our failure to recognize the more general standpoint from which the problem before us appears only as a single link in a chain of related problems. After finding this standpoint, not only is this problem frequently more accessible to our investigation, but at the same time we come into possession of a method which is applicable also to related problems. The introduction of complex paths of integration by Cauchy and of the notion of the IDEALS in number theory by Kummer may serve as examples. This way for finding general methods is certainly the most practicable and the most certain; for he who seeks for methods without having a definite problem in mind seeks for the most part in vain.

In dealing with mathematical problems, specialization plays, as I believe, a still more important part than generalization. Perhaps in most cases where we seek in vain the answer to a question, the cause of the failure lies in the fact that problems simpler and easier than the one in hand have been either not at all or incompletely solved. All depends, then, on finding out these easier problems, and on solving them by means of devices as perfect as possible and of concepts capable of generalization. This rule is one of the most important leers for overcoming mathematical difficulties and it seems to me that it is used almost always, though perhaps unconsciously.

---

[1] Leipzig, 1896.

Occasionally it happens that we seek the solution under insufficient hypotheses or in an incorrect sense, and for this reason do not succeed. The problem then arises: to show the impossibility of the solution under the given hypotheses, or in the sense contemplated. Such proofs of impossibility were effected by the ancients, for instance when they showed that the ratio of the hypotenuse to the side of an isosceles right triangle is irrational. In later mathematics, the question as to the impossibility of certain solutions plays a preëminent part, and we perceive in this way that old and difficult problems, such as the proof of the axiom of parallels, the squaring of the circle, or the solution of equations of the fifth degree by radicals have finally found fully satisfactory and rigorous solutions, although in another sense than that originally intended. It is probably this important fact along with other philosophical reasons that gives rise to the conviction (which every mathematician shares, but which no one has as yet supported by a proof) that every definite mathematical problem must necessarily be susceptible of an exact settlement, either in the form of an actual answer to the question asked, or by the proof of the impossibility of its solution and therewith the necessary failure of all attempts. Take any definite unsolved problem, such as the question as to the irrationality of the Euler-Mascheroni constant C, or the existence of an infinite number of prime numbers of the form $2^n + 1$. However unapproachable these problems may seem to us and however helpless we stand before them, we have, nevertheless, the firm conviction that their solution must follow by a finite number of purely logical processes.

Is this axiom of the solvability of every problem a peculiarity characteristic of mathematical thought alone, or is it possibly a general law inherent in the nature of the mind, that all questions which it asks must be answerable? For in other sciences also one meets old problems which have been settled in a manner most satisfactory and most useful to science by the proof of their impossibility. I instance the problem of perpetual motion. After seeking in vain for the construction of a perpetual motion machine, the relations were investigated which must subsist between the forces of nature if such a machine is to be impossible;[2] and this inverted question led to the discovery of the law of the conservation of energy, which, again, explained the impossibility of perpetual motion in the sense originally intended.

This conviction of the solvability of every mathematical problem is a powerful incentive to the worker. We hear within us the perpetual call: There is the problem. Seek its solution. You can find it by pure reason, for in mathematics there is no *ignorabimus*.

The supply of problems in mathematics is inexhaustible, and as soon as one problem is solved numerous others come forth in its place. Permit me in the following, tentatively as it were, to mention particular definite problems, drawn from various branches of mathematics, from the discussion of which an advancement of science may be expected.

Let us look at the principles of analysis and geometry. The most suggestive and notable achievements of the last century in this field are, as it seems to me, the arithmetical formulation of the concept of the continuum in the works of Cauchy, Bolzano and Cantor, and the discovery of non-euclidean geometry by Gauss, Bolyai,

---

[2]See Helmholtz, "Ueber die Wechselwirkung der Naturkräefte und die darauf bezüglichen neuesten Ermittelungen der Physik"; Vortrag, gehalten in Königsberg, 1854.

and Lobachevsky. I therefore first direct your attention to some problems belonging to these fields.

1. Cantor's problem of the cardinal number of the continuum

Two systems, *i.e.*, two assemblages of ordinary real numbers or points, are said to be (according to Cantor) equivalent or of equal *cardinal number*, if they can be brought into a relation to one another such that to every number of the one assemblage corresponds one and only one definite number of the other. The investigations of Cantor on such assemblages of points suggest a very plausible theorem, which nevertheless, in spite of the most strenuous efforts, no one has succeeded in proving. This is the theorem:

Every system of infinitely many real numbers, *i.e.*, every assemblage of numbers (or points), is either equivalent to the assemblage of natural integers, 1, 2, 3, ... or to the assemblage of all real numbers and therefore to the continuum, that is, to the points of a line; *as regards equivalence there are, therefore, only two assemblages of numbers, the countable assemblage and the continuum.*

From this theorem it would follow at once that the continuum has the next cardinal number beyond that of the countable assemblage; the proof of this theorem would, therefore, form a new bridge between the countable assemblage and the continuum.

Let me mention another very remarkable statement of Cantor's which stands in the closest connection with the theorem mentioned and which, perhaps, offers the key to its proof. Any system of real numbers is said to be ordered, if for every two numbers of the system it is determined which one is the earlier and which the later, and if at the same time this determination is of such a kind that, if a is before b and b is before c, then a always comes before c. The natural arrangement of numbers of a system is defined to be that in which the smaller precedes the larger. But there are, as is easily seen, infinitely many other ways in which the numbers of a system may be arranged.

If we think of a definite arrangement of numbers and select from them a particular system of these numbers, a so-called partial system or assemblage, this partial system will also prove to be ordered. Now Cantor considers a particular kind of ordered assemblage which he designates as a well ordered assemblage and which is characterized in this way, that not only in the assemblage itself but also in every partial assemblage there exists a first number. The system of integers 1, 2, 3, ... in their natural order is evidently a well ordered assemblage. On the other hand the system of all real numbers, *i.e.*, the continuum in its natural order, is evidently not well ordered. For, if we think of the points of a segment of a straight line, with its initial point excluded, as our partial assemblage, it will have no first element.

The question now arises whether the totality of all numbers may not be arranged in another manner so that every partial assemblage may have a first element, *i.e.*, whether the continuum cannot be considered as a well ordered assemblage—a question which Cantor thinks must be answered in the affirmative. It appears to me most desirable to obtain a direct proof of this remarkable statement of Cantor's, perhaps by actually giving an arrangement of numbers such that in every partial system a first number can be pointed out.

2. The compatibility of the arithmetical axioms

When we are engaged in investigating the foundations of a science, we must set up a system of axioms which contains an exact and complete description of the relations subsisting between the elementary ideas of that science. The axioms so set up are at the same time the definitions of those elementary ideas; and no statement within the realm of the science whose foundation we are testing is held to be correct unless it can be derived from those axioms by means of a finite number of logical steps. Upon closer consideration the question arises: *Whether, in any way, certain statements of single axioms depend upon one another, and whether the axioms may not therefore contain certain parts in common, which must be isolated if one wishes to arrive at a system of axioms that shall be altogether independent of one another.*

But above all I wish to designate the following as the most important among the numerous questions which can be asked with regard to the axioms: *To prove that they are not contradictory, that is, that a finite number of logical steps based upon them can never lead to contradictory results.*

In geometry, the proof of the compatibility of the axioms can be effected by constructing a suitable field of numbers, such that analogous relations between the numbers of this field correspond to the geometrical axioms. Any contradiction in the deductions from the geometrical axioms must thereupon be recognizable in the arithmetic of this field of numbers. In this way the desired proof for the compatibility of the geometrical axioms is made to depend upon the theorem of the compatibility of the arithmetical axioms.

On the other hand a direct method is needed for the proof of the compatibility of the arithmetical axioms. The axioms of arithmetic are essentially nothing else than the known rules of calculation, with the addition of the axiom of continuity. I recently collected them[3] and in so doing replaced the axiom of continuity by two simpler axioms, namely, the well-known axiom of Archimedes, and a new axiom essentially as follows: that numbers form a system of things which is capable of no further extension, as long as all the other axioms hold (axiom of completeness). I am convinced that it must be possible to find a direct proof for the compatibility of the arithmetical axioms, by means of a careful study and suitable modification of the known methods of reasoning in the theory of irrational numbers.

To show the significance of the problem from another point of view, I add the following observation: If contradictory attributes be assigned to a concept, I say, that *mathematically the concept does not exist.* So, for example, a real number whose square is -1 does not exist mathematically. But if it can be proved that the attributes assigned to the concept can never lead to a contradiction by the application of a finite number of logical processes, I say that the mathematical existence of the concept (for example, of a number or a function which satisfies certain conditions) is thereby proved. In the case before us, where we are concerned with the axioms of real numbers in arithmetic, the proof of the compatibility of the axioms is at the same time the proof of the mathematical existence of the complete system of real numbers or of the continuum. Indeed, when the proof for the compatibility of the axioms shall be fully accomplished, the doubts which have been expressed occasionally as to the existence of the complete system of real numbers will become totally groundless. The totality of real numbers, *i.e.,* the continuum according to the point of view just indicated, is not the totality of

---

[3] *Jahresbericht der Deutschen Mathematiker-Vereinigung,* vol. 8 (1900), p. 180.

all possible series in decimal fractions, or of all possible laws according to which the elements of a fundamental sequence may proceed. It is rather a system of things whose mutual relations are governed by the axioms set up and for which all propositions, and only those, are true which can be derived from the axioms by a finite number of logical processes. In my opinion, the concept of the continuum is strictly logically tenable in this sense only. It seems to me, indeed, that this corresponds best also to what experience and intuition tell us. The concept of the continuum or even that of the system of all functions exists, then, in exactly the same sense as the system of integral, rational numbers, for example, or as Cantor's higher classes of numbers and cardinal numbers. For I am convinced that the existence of the latter, just as that of the continuum, can be proved in the sense I have described; unlike the system of *all* cardinal numbers or of *all* Cantor's alephs, for which, as may be shown, a system of axioms, compatible in my sense, cannot be set up. Either of these systems is, therefore, according to my terminology, mathematically non-existent.

From the field of the foundations of geometry I should like to mention the following problem:

3. The Equality of the Volumes of Two Tetrahedra of Equal Bases and Equal Altitudes

In two letters to Gerling, Gauss[4] expresses his regret that certain theorems of solid geometry depend upon the method of exhaustion, *i.e.*, in modern phraseology, upon the axiom of continuity (or upon the axiom of Archimedes). Gauss mentions in particular the theorem of Euclid, that triangular pyramids of equal altitudes are to each other as their bases. Now the analogous problem in the plane has been solved.[5] Gerling also succeeded in proving the equality of volume of symmetrical polyhedra by dividing them into congruent parts. Nevertheless, it seems to me probable that a general proof of this kind for the theorem of Euclid just mentioned is impossible, and it should be our task to give a rigorous proof of its impossibility. This would be obtained, as soon as we succeeded in *specifying two tetrahedra of equal bases and equal altitudes which can in no way be split up into congruent tetrahedra, and which cannot be combined with congruent tetrahedra to form two polyhedra which themselves could be split up into congruent tetrahedra.*[6]

4. Problem of the Straight Line as the Shortest Distance Between Two Points

Another problem relating to the foundations of geometry is this: If from among the axioms necessary to establish ordinary euclidean geometry, we exclude the axiom of parallels, or assume it as not satisfied, but retain all other axioms, we obtain, as is well known, the geometry of Lobachevsky (hyperbolic geometry). We may therefore say that this is a geometry standing next to euclidean geometry. If we require further that that axiom be not satisfied whereby, of three points of a straight line, one and only one lies between the other two, we obtain Riemann's (elliptic)

---

[4] Werke, vol. 8, pp. 241 and 244.

[5] Cf., beside earlier literature, Hilbert, Grundlagen der Geometrie, Leipzig, 1899, ch. 4. [Translation by Townsend, Chicago, 1902.]

[6] Since this was written Herr Dehn has succeeded in proving this impossibility. See his note: "Ueber raumgleiche Polyeder", in *Nachrichten d. K. Geselsch. d. Wiss. zu Göttingen*, 1900, and a paper soon to appear in the *Math. Annalen* [vol. 55, pp. 465–478].

geometry, so that this geometry appears to be the next after Lobachevsky's. If we wish to carry out a similar investigation with respect to the axiom of Archimedes, we must look upon this as not satisfied, and we arrive thereby at the non-archimedean geometries which have been investigated by Veronese and myself. The more general question now arises: Whether from other suggestive standpoints geometries may not be devised which, with equal right, stand next to euclidean geometry. Here I should like to direct your attention to a theorem which has, indeed, been employed by many authors as a definition of a straight line, viz., that the straight line is the shortest distance between two points. The essential content of this statement reduces to the theorem of Euclid that in a triangle the sum of two sides is always greater than the third side—a theorem which, as is easily seen, deals solely with elementary concepts, *i.e.*, with such as are derived directly from the axioms, and is therefore more accessible to logical investigation. Euclid proved this theorem, with the help of the theorem of the exterior angle, on the basis of the congruence theorems. Now it is readily shown that this theorem of Euclid cannot be proved solely on the basis of those congruence theorems which relate to the application of segments and angles, but that one of the theorems on the congruence of triangles is necessary. We are asking, then, for a geometry in which all the axioms of ordinary euclidean geometry hold, and in particular all the congruence axioms except the one of the congruence of triangles (or all except the theorem of the equality of the base angles in the isosceles triangle), and in which, besides, the proposition that in every triangle the sum of two sides is greater than the third is assumed as a particular axiom.

One finds that such a geometry really exists and is no other than that which Minkowski constructed in his book, Geometrie der Zahlen,[7] and made the basis of his arithmetical investigations. Minkowski's is therefore also a geometry standing next to the ordinary euclidean geometry; it is essentially characterized by the following stipulations:

1. The points which are at equal distances from a fixed point O lie on a convex closed surface of the ordinary euclidean space with O as a center.

2. Two segments are said to be equal when one can be carried into the other by a translation of the ordinary euclidean space.

In Minkowski's geometry the axiom of parallels also holds. By studying the theorem of the straight line as the shortest distance between two points, I arrived[8] at a geometry in which the parallel axiom does not hold, while all other axioms of Minkowski's geometry are satisfied. The theorem of the straight line as the shortest distance between two points and the essentially equivalent theorem of Euclid about the sides of a triangle, play an important part not only in number theory but also in the theory of surfaces and in the calculus of variations. For this reason, and because I believe that the thorough investigation of the conditions for the validity of this theorem will throw a new light upon the idea of distance, as well as upon other elementary ideas, *e.g.*, upon the idea of the plane, and the possibility of its definition by means of the idea of the straight line, *the construction and systematic treatment of the geometries here possible seem to me desirable.*

---

[7] Leipzig, 1896.
[8] *Math. Annalen*, vol. 46, p. 91.

5. Lie's concept of a continuous group of transformations without the assumption of the differentiability of the functions defining the group

It is well known that Lie, with the aid of the concept of continuous groups of transformations, has set up a system of geometrical axioms and, from the standpoint of his theory of groups, has proved that this system of axioms suffices for geometry. But since Lie assumes, in the very foundation of his theory, that the functions defining his group can be differentiated, it remains undecided in Lie's development, whether the assumption of the differentiability in connection with the question as to the axioms of geometry is actually unavoidable, or whether it may not appear rather as a consequence of the group concept and the other geometrical axioms. This consideration, as well as certain other problems in connection with the arithmetical axioms, brings before us the more general question: *How far Lie's concept of continuous groups of transformations is approachable in our investigations without the assumption of the differentiability of the functions.*

Lie defines a finite continuous group of transformations as a system of transformations

$$x_i' = f_i(x_1, \ldots, x_n; a_1, \ldots, a_r) \qquad (i = 1, \ldots, n)$$

having the property that any two arbitrarily chosen transformations of the system, as

$$x_i' = f_i(x_1, \ldots, x_n; a_1, \ldots, a_r),$$
$$x_i'' = f_i(x_1', \ldots, x_n'; b_1, \ldots, b_r),$$

applied successively result in a transformation which also belongs to the system, and which is therefore expressible in the form

$$x_i'' = f_i\{f_1(x, a), \ldots, f_n(x, a); b_1, \ldots, b_r\} = f_i(x_1, \ldots, x_n; c_1, \ldots, c_r),$$

where $c_1, \ldots, c_r$ are certain functions of $a_1, \ldots, a_r$ and $b_1, \ldots, b_r$. The group property thus finds its full expression in a system of functional equations and of itself imposes no additional restrictions upon the functions $f_1, \ldots, f_n; c_1, \ldots, c_r$. Yet Lie's further treatment of these functional equations, viz., the derivation of the well-known fundamental differential equations, assumes necessarily the continuity and differentiability of the functions defining the group.

As regards continuity: this postulate will certainly be retained for the present— if only with a view to the geometrical and arithmetical applications, in which the continuity of the functions in question appears as a consequence of the axiom of continuity. On the other hand the differentiability of the functions defining the group contains a postulate which, in the geometrical axioms, can be expressed only in a rather forced and complicated manner. Hence there arises the question whether, through the introduction of suitable new variables and parameters, the group can always be transformed into one whose defining functions are differentiable; or whether, at least with the help of certain simple assumptions, a transformation is possible into groups admitting Lie's methods. A reduction to analytic groups is, according to a theorem announced by Lie[9] but first proved by

---

[9]Lie-Engel, Theorie der Transformationsgruppen, vol. 3, Leipzig, 1893, §§82, 144.

Schur,[10] always possible when the group is transitive and the existence of the first and certain second derivatives of the functions defining the group is assumed.

For infinite groups the investigation of the corresponding question is, I believe, also of interest. Moreover we are thus led to the wide and interesting field of functional equations which have been heretofore investigated usually only under the assumption of the differentiability of the functions involved. In particular the functional equations treated by Abel[11] with so much ingenuity, the difference equations, and other equations occurring in the literature of mathematics, do not directly involve anything which necessitates the requirement of the differentiability of the accompanying functions. In the search for certain existence proofs in the calculus of variations I came directly upon the problem: To prove the differentiability of the function under consideration from the existence of a difference equation. In all these cases, then, the problem arises: *In how far are the assertions which we can make in the case of differentiable functions true under proper modifications without this assumption?*

It may be further remarked that H. Minkowski in his above-mentioned Geometrie der Zahlen starts with the functional equation

$$f(x_1 + y_1, \ldots, x_n + y_n) \le f(x_1, \ldots, x_n) + f(y_1, \ldots, y_n)$$

and from this actually succeeds in proving the existence of certain differential quotients for the function in question.

On the other hand I wish to emphasize the fact that there certainly exist analytical functional equations whose sole solutions are non-differentiable functions. For example a uniform continuous non-differentiable function $\varphi(x)$ can be constructed which represents the only solution of the two functional equations

$$\varphi(x + \alpha) - \varphi(x) = f(x), \qquad \varphi(x + \beta) - \varphi(x) = 0,$$

where α and β are two real numbers, and $f(x)$ denotes, for all the real values of x, a regular analytic uniform function. Such functions are obtained in the simplest manner by means of trigonometrical series by a process similar to that used by Borel (according to a recent announcement of Picard)[12] for the construction of a doubly periodic, non-analytic solution of a certain analytic partial differential equation.

6. MATHEMATICAL TREATMENT OF THE AXIOMS OF PHYSICS

The investigations on the foundations of geometry suggest the problem: *To treat in the same manner, by means of axioms, those physical sciences in which mathematics plays an important part; in the first rank are the theory of probabilities and mechanics.*

As to the axioms of the theory of probabilities,[13] it seems to me desirable that their logical investigation should be accompanied by a rigorous and satisfactory development of the method of mean values in mathematical physics, and in particular in the kinetic theory of gases.

---

[10] "Ueber den analytischen Charakter der eine endliche Kontinuierliche Transformationsgruppen darstellenden Funktionen", *Math. Annalen*, vol. 41.

[11] Werke, vol. 1, pp. 1, 61, 389.

[12] "Quelques théories fondamentales dans l'analyse mathématique", Conférences faites à Clark University, *Revue générale des Sciences*, 1900, p. 22.

[13] Cf. Bohlmann, "Ueber Versicherungsmathematik", from the collection: Klein and Riecke, Ueber angewandte Mathematik und Physik, Leipzig, 1900.

Important investigations by physicists on the foundations of mechanics are at hand; I refer to the writings of Mach,[14] Hertz,[15] Boltzmann[16] and Volkmann.[17] It is therefore very desirable that the discussion of the foundations of mechanics be taken up by mathematicians also. Thus Boltzmann's work on the principles of mechanics suggests the problem of developing mathematically the limiting processes, there merely indicated, which lead from the atomistic view to the laws of motion of continua. Conversely one might try to derive the laws of the motion of rigid bodies by a limiting process from a system of axioms depending upon the idea of continuously varying conditions of a material filling all space continuously, these conditions being defined by parameters. For the question as to the equivalence of different systems of axioms is always of great theoretical interest.

If geometry is to serve as a model for the treatment of physical axioms, we shall try first by a small number of axioms to include as large a class as possible of physical phenomena, and then by adjoining new axioms to arrive gradually at the more special theories. At the same time Lie's principle of subdivision can perhaps be derived from profound theory of infinite transformation groups. The mathematician will have also to take account not only of those theories coming near to reality, but also, as in geometry, of all logically possible theories. He must be always alert to obtain a complete survey of all conclusions derivable from the system of axioms assumed.

Further, the mathematician has the duty to test exactly in each instance whether the new axioms are compatible with the previous ones. The physicist, as his theories develop, often finds himself forced by the results of his experiments to make new hypotheses, while he depends, with respect to the compatibility of the new hypotheses with the old axioms, solely upon these experiments or upon a certain physical intuition, a practice which in the rigorously logical building up of a theory is not admissible. The desired proof of the compatibility of all assumptions seems to me also of importance, because the effort to obtain each proof always forces us most effectually to an exact formulation of the axioms.

So far we have considered only questions concerning the foundations of the mathematical sciences. Indeed, the study of the foundations of a science is always particularly attractive, and the testing of these foundations will always be among the foremost problems of the investigator. Weierstrass once said, "The final object always to be kept in mind is to arrive at a correct understanding of the foundations of the science....But to make any progress in the sciences the study of particular problems is, of course, indispensable." In fact, a thorough understanding of its special theories is necessary to the successful treatment of the foundations of the science. Only that architect is in the position to lay a sure foundation for a structure who knows its purpose thoroughly and in detail. So we turn now to the special problems of the separate branches of mathematics and consider first arithmetic and algebra.

---

[14] Die Mechanik in ihrer Entwickelung, Leipzig, 4th edition, 1901.

[15] Die Prinzipien der Mechanik, Leipzig, 1894.

[16] Vorlesungen über die Principe der Mechanik, Leipzig, 1897.

[17] Einführung in das Studium der theoretischen Physik, Leipzig, 1900.

7. IRRATIONALITY AND TRANSCENDENCE OF CERTAIN NUMBERS

Hermite's arithmetical theorems on the exponential function and their extension by Lindemann are certain of the admiration of all generations of mathematicians. Thus the task at once presents itself to penetrate further along the path here entered, as A. Hurwitz has already done in two interesting papers,[18] "Ueber arithmetische Eigenschaften gewisser transzendenter Funktionen". I should like, therefore, to sketch a class of problems which, in my opinion, should be attacked as here next in order. That certain special transcendental functions, important in analysis, take algebraic values for certain algebraic arguments, seems to us particularly remarkable and worthy of thorough investigation. Indeed, we expect transcendental functions to assume, in general, transcendental values for even algebraic arguments; and, although it is well known that there exist integral transcendental functions which even have rational values for all algebraic arguments, we shall still consider it highly probable that the exponential functions $e^{i\pi z}$, for example, which evidently has algebraic values for all rational arguments z, will on the other hand always take transcendental values for irrational algebraic values of the argument z. We can also give this statement a geometrical form, as follows:

If, in an isosceles triangle, the ratio of the base angle to the angle at the vertex be algebraic but not rational, the ratio between base and side is always transcendental.

In spite of the simplicity of this statement and of its similarity to the problems solved by Hermite and Lindemann, I consider the proof of this theorem very difficult; as also the proof that

The expression α^β, for an algebraic base α and an irrational algebraic exponent β, e.g., the number $2^{\sqrt{2}}$ or $e^\pi = i^{-2i}$, always represents a transcendental or at least an irrational number.

It is certain that the solution of these and similar problems must lead us to entirely new methods and to a new insight into the nature of special irrational and transcendental numbers.

8. PROBLEMS OF PRIME NUMBERS

Essential progress in the theory of the distribution of prime numbers has lately been made by Hadamard, de la Vallée-Poussin, Von Mangoldt and others. For the complete solution, however, of the problems set us by Riemann's paper "Ueber die Anzahl der Primzahlen unter einer gegebenen Grösse", it still remains to prove the correctness of an exceedingly important statement of Riemann, viz., *that the zero points of the function $\zeta(s)$ defined by the series*

$$\zeta(s) = 1 + \frac{1}{2^s} + \frac{1}{3^s} + \frac{1}{4^s} + \cdots$$

all have the real part $\frac{1}{2}$, except the well-known negative integral real zeros. As soon as this proof has been successfully established, the next problem would consist in testing more exactly Riemann's infinite series for the number of primes below a given number and, especially, *to decide whether the difference between the number of primes below a number x and the integral logarithm of x does in fact become infinite of an order not greater than $\frac{1}{2}$ in x.*[19] Further, we should determine whether

---

[18] *Math. Annalen*, vols. 22, 32 (1883, 1888).

[19] Cf. an article by H. von Koch, which is soon to appear in the *Math. Annalen* [Vol. 55, p. 441].

the occasional condensation of prime numbers which has been noticed in counting primes is really due to those terms of Riemann's formula which depend upon the first complex zeros of the function $\zeta(s)$.

After an exhaustive discussion of Riemann's prime number formula, perhaps we may sometime be in a position to attempt the rigorous solution of Goldbach's problem,[20] viz., whether every integer is expressible as the sum of two positive prime numbers; and further to attack the well-known question, whether there are an infinite number of pairs of prime numbers with the difference 2, or even the more general problem, whether the linear diophantine equation

$$ax + by + c = 0$$

(with given integral coefficients each prime to the others) is always solvable in prime numbers x and y.

But the following problem seems to me of no less interest and perhaps of still wider range: *To apply the results obtained for the distribution of rational prime numbers to the theory of the distribution of ideal primes in a given number-field* k—a problem which looks toward the study of the function $\zeta_k(s)$ belonging to the field and defined by the series

$$\zeta_k(s) = \sum \frac{1}{n(j)^s},$$

where the sum extends over all ideals j of the given realm k, and $n(j)$ denotes the norm of the ideal j.

I may mention three more special problems in number theory: one on the laws of reciprocity, one on diophantine equations, and a third from the realm of quadratic forms.

9. Proof of the most general law of reciprocity in any number field

For any field of numbers the law of reciprocity is to be proved for the residues of the lth *power, when* l *denotes an odd prime, and further when* l *is a power of 2 or a power of an odd prime.*

The law, as well as the means essential to its proof, will, I believe, result by suitably generalizing the theory of the field of the lth roots of unity,[21] developed by me, and my theory of relative quadratic fields.[22]

10. Determination of the solvability of a diophantine equation

Given a diophantine equation with any number of unknown quantities and with rational integral numerical coefficients: *To devise a process according to which it can be determined by a finite number of operations whether the equation is solvable in rational integers.*

---

[20]Cf. P. Stäckel: "Über Goldbach's empirisches Theorem", *Nachrichten d. K. Ges. d. Wiss. zu Göttingen*, 1896, and Landau, *ibid.*, 1900.

[21]*Jahresber. d. Deutschen Math.-Vereinigung*, "Ueber die Theorie der algebraischen Zahlkörper", vol. 4 (1897), Part V.

[22]*Math. Annalen*, vol. 51 and *Nachrichten d. K. Ges. d. Wiss. zu Göttingen*, 1898.

11. Quadratic forms with any algebraic numerical coefficients

Our present knowledge of the theory of quadratic number fields[23] puts us in a position *to attack successfully the theory of quadratic forms with any number of variables and with any algebraic numerical coefficients*. This leads in particular to the interesting problem: to solve a given quadratic equation with algebraic numerical coefficients in any number of variables by integral or fractional numbers belonging to the algebraic realm of rationality determined by the coefficients.

The following important problem may form a transition to algebra and the theory of functions:

12. Extension of Kronecker's theorem on abelian fields to any algebraic realm of rationality

The theorem that every abelian number field arises from the realm of rational numbers by the composition of fields of roots of unity is due to Kronecker. This fundamental theorem in the theory of integral equations contains two statements, namely:

First. It answers the question as to the number and existence of those equations which have a given degree, a given abelian group and a given discriminant with respect to the realm of rational numbers.

Second. It states that the roots of such equations form a realm of algebraic numbers which coincides with the realm obtained by assigning to the argument z in the exponential function $e^{i\pi z}$ all rational numerical values in succession.

The first statement is concerned with the question of the determination of certain algebraic numbers by their groups and their branching. This question corresponds, therefore, to the known problem of the determination of algebraic functions corresponding to given Riemann surfaces. The second statement furnishes the required numbers by transcendental means, namely, by the exponential function $e^{i\pi z}$.

Since the realm of the imaginary quadratic number fields is the simplest after the realm of rational numbers, the problem arises, to extend Kronecker's theorem to this case. Kronecker himself has made the assertion that the abelian equations in the realm of a quadratic field are given by the equations of transformation of elliptic functions with singular moduli, so that the elliptic function assumes here the same rôle as the exponential function in the former case. The proof of Kronecker's conjecture has not yet been furnished; but I believe that it must be obtainable without very great difficulty on the basis of the theory of complex multiplication developed by H. Weber[24] with the help of the purely arithmetical theorems on class fields which I have established.

Finally, the extension of Kronecker's theorem to the case that, *in place of the realm of rational numbers or of the imaginary quadratic field, any algebraic field*

---

[23]Hilbert, "Ueber den Dirichlet'schen biquadratischen Zahlenkörper", *Math. Annalen*, vol. 45; "Ueber die Theorie der relativquadratischen Zahlkörper", *Jahresber. d. Deutschen Mathematiker-Vereinigung*, 1897, and *Math. Annalen*, vol. 51; "Ueber die Theorie der relativ-Abelschen Körper", *Nachrichten d. K. Ges. d. Wiss. zu Göttingen*, 1898; Grundlagen der Geometrie, Leipzig, 1899, Chap. VIII, §83 [Translation by Townsend, Chicago, 1902]. Cf. also the dissertation of G. Rückle, Göttingen, 1901.

[24]Elliptische Functionen und algebraische Zahlen, Braunschweig, 1891.

whatever is laid down as realm of rationality, seems to me of the greatest importance. I regard this problem as one of the most profound and far-reaching in the theory of numbers and of functions.

The problem is found to be accessible from many standpoints. I regard as the most important key to the arithmetical part of this problem the general law of reciprocity for residues of lth powers within any given number field.

As to the function-theoretical part of the problem, the investigator in this attractive region will be guided by the remarkable analogies which are noticeable between the theory of algebraic functions of one variable and the theory of algebraic numbers. Hensel[25] has proposed and investigated the analogue in the theory of algebraic numbers to the development in power series of an algebraic function; and Landsberg[26] has treated the analogue of the Riemann-Roch theorem. The analogy between the deficiency of a Riemann surface and that of the class number of a field of numbers is also evident. Consider a Riemann surface of deficiency $p = 1$ (to touch on the simplest case only) and on the other hand a number field of class $h = 2$. To the proof of the existence of an integral everywhere finite on the Riemann surface, corresponds the proof of the existence of an integer α in the number field such that the number $\sqrt{\alpha}$ represents a quadratic field, relatively unbranched with respect to the fundamental field. In the theory of algebraic functions, the method of boundary values (*Randwerthaufgabe*) serves, as is well known, for the proof of Riemann's existence theorem. In the theory of number fields also, the proof of the existence of just this number α offers the greatest difficulty. This proof succeeds with indispensable assistance from the theorem that in the number field there are always prime ideals corresponding to given residual properties. This latter fact is therefore the analogue in number theory to the problem of boundary values.

The equation of Abel's theorem in the theory of algebraic functions expresses, as is well known, the necessary and sufficient condition that the points in question on the Riemann surface are the zero points of an algebraic function belonging to the surface. The exact analogue of Abel's theorem, in the theory of the number field of class $h = 2$, is the equation of the law of quadratic reciprocity[27]

$$\left(\frac{\alpha}{j} \right) = +1,$$

which declares that the ideal j is then and only then a principal ideal of the number field when the quadratic residue of the number α with respect to the ideal j is positive.

It will be seen that in the problem just sketched the three fundamental branches of mathematics, number theory, algebra and function theory, come into closest touch with one another, and I am certain that the theory of analytical functions of several variables in particular would be notably enriched if one should succeed *in finding and discussing those functions which play the part for any algebraic number field corresponding to that of the exponential function in the field of rational numbers and of the elliptic modular functions in the imaginary quadratic number field.*

---

[25] *Jahresber. d. Deutschen Math.-Vereinigung,* vol. 6, and an article soon to appear in the *Math. Annalen* [Vol. 55, p. 301]: "Ueber die Entwickelung der algebraischen Zahlen in Potenzreihen".

[26] *Math. Annalen,* vol. 50 (1898).

[27] Cf. Hilbert, "Ueber die Theorie der relativ-Abelschen Zahlkörper", *Gött. Nachrichten,* 1898.

Passing to algebra, I shall mention a problem from the theory of equations and one to which the theory of algebraic invariants has led me.

13. IMPOSSIBILITY OF THE SOLUTION OF THE GENERAL EQUATION OF THE 7TH DEGREE BY MEANS OF FUNCTIONS OF ONLY TWO ARGUMENTS

Nomography[28] deals with the problem: to solve equations by means of drawings of families of curves depending on an arbitrary parameter. It is seen at once that every root of an equation whose coefficients depend upon only two parameters, that is, every function of two independent variables, can be represented in manifold ways according to the principle lying at the foundation of nomography. Further, a large class of functions of three or more variables can evidently be represented by this principle alone without the use of variable elements, namely all those which can be generated by forming first a function of two arguments, then equating each of these arguments to a function of two arguments, next replacing each of those arguments in their turn by a function of two arguments, and so on, regarding as admissible any finite number of insertions of functions of two arguments. So, for example, every rational function of any number of arguments belongs to this class of functions constructed by nomographic tables; for it can be generated by the processes of addition, subtraction, multiplication and division and each of these processes produces a function of only two arguments. One sees easily that the roots of all equations which are solvable by radicals in the natural realm of rationality belong to this class of functions; for here the extraction of roots is adjoined to the four arithmetical operations and this, indeed, presents a function of one argument only. Likewise the general equations of the 5th and 6th degrees are solvable by suitable nomographic tables; for, by means of Tschirnhausen transformations, which require only extraction of roots, they can be reduced to a form where the coefficients depend upon two parameters only.

Now it is probable that the root of the equation of the seventh degree is a function of its coefficients which does not belong to this class of functions capable of nomographic construction, *i.e.*, that it cannot be constructed by a finite number of insertions of functions of two arguments. In order to prove this, the proof would be necessary *that the equation of the seventh degree* $f^7 + xf^3 + yf^2 + zf + 1 = 0$ *is not solvable with the help of any continuous functions of only two arguments.* I may be allowed to add that I have satisfied myself by a rigorous process that there exist analytical functions of three arguments x, y, z which cannot be obtained by a finite chain of functions of only two arguments.

By employing auxiliary movable elements, nomography succeeds in constructing functions of more than two arguments, as d'Ocagne has recently proved in the case of the equation of the 7th degree.[29]

---

[28] d'Ocagne, Traité de Nomographie, Paris, 1899.

[29] "Sur la résolution nomographique de l'équation du septième degré", *Comptes rendus*, Paris, 1900.

14. Proof of the Finiteness of Certain Complete
Systems of Functions

In the theory of algebraic invariants, questions as to the finiteness of complete systems of forms deserve, as it seems to me, particular interest. L. Maurer[30] has lately succeeded in extending the theorems on finiteness in invariant theory proved by P. Gordan and myself, to the case where, instead of the general projective group, any subgroup is chosen as the basis for the definition of invariants.

An important step in this direction had been taken already by A. Hurwitz,[31] who, by an ingenious process, succeeded in effecting the proof, in its entire generality, of the finiteness of the system of orthogonal invariants of an arbitrary ground form.

The study of the question as to the finiteness of invariants has led me to a simple problem which includes that question as a particular case and whose solution probably requires a decidedly more minutely detailed study of the theory of elimination and of Kronecker's algebraic modular systems than has yet been made.

Let a number m of integral rational functions $X_1, X_2, \ldots, X_m$ for the n variables $x_1, x_2, \ldots, x_n$ be given,

$$(S) \qquad \begin{aligned} X_1 &= f_1(x_1, \ldots, x_n), \\ X_2 &= f_2(x_1, \ldots, x_n), \\ &\,\, \ldots\ldots\ldots\ldots\ldots\ldots\ldots \\ X_m &= f_m(x_1, \ldots, x_n). \end{aligned}$$

Every rational integral combination of $X_1, \ldots, X_m$ must evidently always become, after substitution of the above expressions, a rational integral function of $x_1, \ldots, x_n$. Nevertheless, there may well be rational fractional functions of $X_1, \ldots, X_m$ which, by the operation of the substitution S, become integral functions in $x_1, \ldots, x_m$, which becomes integral in $x_1, \ldots, x_n$ after the application of the substitution S, I propose to call a *relatively integral* function of $X_1, \ldots, X_m$. Every integral function of $X_1, \ldots, X_m$ is evidently also relatively integral; further the sum, difference and product of relative integral functions are themselves relatively integral.

The resulting problem is now to decide whether it is always possible *to find a finite system of relatively integral function $X_1, \ldots, X_m$ by which every other relatively integral function of $X_1, \ldots, X_m$ may be expressed rationally and integrally.*

We can formulate the problem still more simply if we introduce the idea of a finite field of integrality. By a finite field of integrality I mean a system of functions from which a finite number of functions can be chosen, in terms of which all other functions of the system are rationally and integrally expressible. Our problem amounts, then, to this: to show that all relatively integral functions of any given domain of rationality always constitute a finite field of integrality.

It naturally occurs to us also to refine the problem by restrictions drawn from number theory, by assuming the coefficients of the given functions $f_1, \ldots, f_m$ to be integers and including among the relatively integral functions of $X_1, \ldots, X_m$ only such rational functions of these arguments as become, by the application of the substitutions S, rational integral functions of $x_1, \ldots, x_n$ with rational integral coefficients.

---

[30] Cf. *Sitzungsber. d. K. Acad. d. Wiss. zu München*, 1899, and an article about to appear in the *Math. Annalen*.

[31] "Ueber die Erzeugung der Invarianten durch Integration", *Nachrichten d. K. Gesellschaft d. Wiss. zu Göttingen*, 1897.

The following is a simple particular case of this refined problem: Let m integral rational functions $X_1, \ldots, X_m$ of one variable x with integral rational coefficients, and a prime number p be given. Consider the system of those integral rational functions of x which can be expressed in the form

$$\frac{G(X_1, \ldots, X_m)}{p^h},$$

where G is a rational integral function of the arguments $X_1, \ldots, X_m$ and p^h is any power of the prime number p. Earlier investigations of mine[32] show immediately that all such expressions for a fixed exponent h form a finite domain of integrality. But the question here is whether the same is true for all exponents h, i.e., whether a finite number of such expressions can be chosen by means of which for every exponent h every other expression of that form is integrally and rationally expressible.

From the boundary region between algebra and geometry, I will mention two problems. The one concerns enumerative geometry and the other the topology of algebraic curves and surfaces.

15. RIGOROUS FOUNDATION OF SCHUBERT'S ENUMERATIVE CALCULUS

The problem consists in this: *To establish rigorously and with an exact determination of the limits of their validity those geometrical numbers which Schubert[33] especially has determined on the basis of the so-called principle of special position, or conservation of number, by means of the enumerative calculus developed by him.*

Although the algebra of to-day guarantees, in principle, the possibility of carrying out the processes of elimination, yet for the proof of the theorems of enumerative geometry decidedly more is requisite, namely, the actual carrying out of the process of elimination in the case of equations of special form in such a way that the degree of the final equations and the multiplicity of their solutions may be foreseen.

16. PROBLEM OF THE TOPOLOGY OF ALGEBRAIC CURVES AND SURFACES

The maximum number of closed and separate branches which a plane algebraic curve of the nth order can have has been determined by Harnack.[34] There arises the further question as to the relative position of the branches in the plane. As to curves of the 6th order, I have satisfied myself—by a complicated process, it is true—that of the eleven branches which they can have according to Harnack, by no means all can lie external to one another, but that one branch must exist in whose interior one branch and in whose exterior nine branches lie, or inversely. *A thorough investigation of the relative position of the separate branches when their number is the maximum seems to me to be of very great interest, and not less so the corresponding investigation as to the number, form, and position of the sheets of an algebraic surface in space.* Till now, indeed, it is not even known what is the maximum number of sheets which a surface of the 4th order in three dimensional space can really have.[35]

---

[32] *Math. Annalen*, vol. 36 (1890), p. 485.

[33] Kalkül der abzählenden Geometrie, Leipzig, 1879.

[34] *Math. Annalen*, vol. 10.

[35] Cf. Rohn, "Flächen vierter Ordnung", Preisschriften der Fürstlich Jablonowskischen Gesellschaft, Leipzig, 1886.

In connection with this purely algebraic problem, I wish to bring forward a question which, it seems to me, may be attacked by the same method of continuous variation of coefficients, and whose answer is of corresponding value for the topology of families of curves defined by differential equations. This is the question as to the maximum number and position of Poincaré's boundary cycles (cycles limites) for a differential equation of the first order and degree of the form

$$\frac{dy}{dx} = \frac{Y}{X},$$

where X and Y are rational integral functions of the nth degree in x and y. Written homogeneously, this is

$$X\left(y\frac{dz}{dt} - z\frac{dy}{dt}\right) + Y\left(z\frac{dx}{dt} - x\frac{dz}{dt}\right) + Z\left(x\frac{dy}{dt} - y\frac{dx}{dt}\right) = 0,$$

where X, Y, and Z are rational integral homogeneous functions of the nth degree in x, y, z, and the latter are to be determined as functions of the parameter t.

17. EXPRESSION OF DEFINITE FORMS BY SQUARES

A rational integral function or form in any number of variables with real coefficients such that it becomes negative for no real values of these variables, is said to be *definite*. The system of all definite forms is invariant with respect to the operations of addition and multiplication, but the quotient of two definite forms—in case it should be an integral function of the variables—is also a definite form. The square of any form is evidently always a definite form. But since, as I have shown,[36] not every definite form can be compounded by addition from squares of forms, the question arises—which I have answered affirmatively for ternary forms[37]—whether every definite form may not be expressed as a quotient of sums of squares of forms. At the same time it is desirable, for certain questions as to the possibility of certain geometrical constructions, to know whether the coefficients of the forms to be used in the expression may always be taken from the realm of rationality given by the coefficients of the form represented.[38]

I mention one more geometrical problem:

18. BUILDING UP OF SPACE FROM CONGRUENT POLYHEDRA

If we enquire for those groups of motions in the plane for which a fundamental region exists, we obtain various answers, according as the plane considered is Riemann's (elliptic), Euclid's, or Lobachevsky's (hyperbolic). In the case of the elliptic plane there is a finite number of essentially different kinds of fundamental regions, and a finite number of congruent regions suffices for a complete covering of the whole plane; the group consists indeed of a finite number of motions only. In the case of the hyperbolic plane there is an infinite number of essentially different kinds of fundamental regions, namely, the well-known Poincaré polygons. For the complete covering of the plane an infinite number of congruent regions is necessary. The case of Euclid's plane stands between these; for in this case there is only a

---

[36] *Math. Annalen*, vol. 32.

[37] *Acta Mathematica*, vol. 17.

[38] Cf. Hilbert: Grundlagen der Geometrie, Leipzig, 1899, Chap. 7 and in particular §38.

finite number of essentially different kinds of groups of motions with fundamental regions, but for a complete covering of the whole plane an infinite number of congruent regions is necessary.

Exactly the corresponding facts are found in space of three dimensions. The fact of the finiteness of the groups of motions in elliptic space is an immediate consequence of a fundamental theorem of C. Jordan,[39] whereby the number of essentially different kinds of finite groups of linear substitutions in n variables does not surpass a certain finite limit dependent upon n. The groups of motions with fundamental regions in hyperbolic space have been investigated by Fricke and Klein in the lectures on the theory of automorphic functions,[40] and finally Fedorov,[41] Schoenflies[42] and lately Rohn[43] have given the proof that there are, in euclidean space, only a finite number of essentially different kinds of groups of motions with a fundamental region. Now, while the results and methods of proof applicable to elliptic and hyperbolic space hold directly for n-dimensional space also, the generalization of the theorem for euclidean space seems to offer decided difficulties. The investigation of the following question is therefore desirable: *Is there in n-dimensional euclidean space also only a finite number of essentially different kinds of groups of motions with a fundamental region?*

A fundamental region of each group of motions, together with the congruent regions arising from the group, evidently fills up space completely. The question arises: *Whether polyhedra also exist which do not appear as fundamental regions of groups of motions, by means of which nevertheless by a suitable juxtaposition of congruent copies a complete filling up of all space is possible.* I point out the following question, related to the preceding one, and important to number theory and perhaps sometimes useful to physics and chemistry: How can one arrange most densely in space an infinite number of equal solids of given form, *e.g.*, spheres with given radii or regular tetrahedra with given edges (or in prescribed position), that is, how can one so fit them together that the ratio of the filled to the unfilled space may be as great as possible?

If we look over the development of the theory of functions in the last century, we notice above all the fundamental importance of that class of functions which we now designate as analytic functions—a class of functions which will probably stand permanently in the center of mathematical interest.

There are many different standpoints from which we might choose, out of the totality of all conceivable functions, extensive classes worthy of a particularly thorough investigation. Consider, for example, *the class of functions characterized by ordinary or partial algebraic differential equations*. It should be observed that this class does not contain the functions that arise in number theory and whose investigation is of the greatest importance. For example, the before-mentioned function $\zeta(s)$ satisfies no algebraic differential equation, as is easily seen with the help of the well-known relation between $\zeta(s)$ and $\zeta(1-s)$, if one refers to the theorem proved by

---

[39] *Crelle's Journal*, vol. 84 (1878), and *Atti d. Reale Acad. di Napoli*, 1880.
[40] Leipzig, 1897. Cf. especially Abschnitt I, Chapters 2 and 3.
[41] Symmetrie der regelmässigen Systeme von Figuren, 1890.
[42] Krystallsysteme und Krystallstruktur, Leipzig, 1891.
[43] *Math. Annalen*, vol. 53.

Hölder,[44] that the function $\Gamma(x)$ satisfies no algebraic differential equation. Again, the function of the two variables s and x defined by the infinite series

$$\zeta(s, x) = x + \frac{x^2}{2^s} + \frac{x^3}{3^s} + \frac{x^4}{4^s} + \cdots,$$

which stands in close relation with the function $\zeta(s)$, probably satisfies no algebraic partial differential equation. In the investigation of this question the functional equation

$$x \frac{\partial \zeta(s, x)}{\partial x} = \zeta(s - 1, x)$$

will have to be used.

If, on the other hand, we are led by arithmetical or geometrical reasons to consider the class of all those functions which are continuous and indefinitely differentiable, we should be obliged in its investigation to dispense with that pliant instrument, the power series, and with the circumstance that the function is fully determined by the assignment of values in any region, however small. While, therefore, the former limitation of the field of functions was too narrow, the latter seems to me too wide.

The idea of the analytic function on the other hand includes the whole wealth of functions most important to science, whether they have their origin in number theory, in the theory of differential equations or of algebraic functional equations, whether they arise in geometry or in mathematical physics; and, therefore, in the entire realm of functions, the analytic function justly holds undisputed supremacy.

19. ARE THE SOLUTIONS OF REGULAR PROBLEMS IN THE CALCULUS OF VARIATIONS ALWAYS NECESSARILY ANALYTIC?

One of the most remarkable facts in the elements of the theory of analytic functions appears to me to be this: That there exist partial differential equations whose integrals are all of necessity analytic functions of the independent variables, that is, in short, equations susceptible of none but analytic solutions. The best known partial differential equations of this kind are the potential equation

$$\frac{\partial^2 f}{\partial x^2} + \frac{\partial^2 f}{\partial y^2} = 0$$

and certain linear differential equations investigated by Picard;[45] also the equation

$$\frac{\partial^2 f}{\partial x^2} + \frac{\partial^2 f}{\partial y^2} = e^f,$$

the partial differential equation of minimal surfaces, and others. Most of these partial differential equations have the common characteristic of being the lagrangian differential equations of certain problems of variation, viz., of such problems of variation

$$\int \int F(p, q, z; x, y) \, dx \, dy = \text{minimum}$$

$$\left[p = \frac{\partial z}{\partial x}, q = \frac{\partial z}{\partial y} \right],$$

---

[44] *Math. Annalen*, vol. 28.

[45] *Jour. de l'Ecole Polytech.*, 1890.

as satisfy, for all values of the arguments which fall within the range of discussion, the inequality

$$\frac{\partial^2 F}{\partial p^2} \cdot \frac{\partial^2 F}{\partial q^2} - \left(\frac{\partial^2 F}{\partial p \partial q}\right)^2 > 0,$$

F itself being an analytic function. We shall call this sort of problem a *regular* variation problem. It is chiefly the regular variation problems that play a rôle in geometry, in mechanics, and in mathematical physics; and the question naturally arises, whether all solutions of regular variation problems must necessarily be analytic functions. In other words, *does every lagrangian partial differential equation of a regular variation problem have the property of admitting analytic integrals exclusively?* And is this the case even when the function is constrained to assume, as, *e.g.*, in Dirichlet's problem on the potential function, boundary values which are continuous, but not analytic?

I may add that there exist surfaces of constant *negative* gaussian curvature which are representable by functions that are continuous and possess indeed all the derivatives, and yet are not analytic; while on the other hand it is probable that every surface whose gaussian curvature is constant and *positive* is necessarily an analytic surface. And we know that the surfaces of positive constant curvature are most closely related to this regular variation problem: To pass through a closed curve in space a surface of minimal area which shall inclose, in connection with a fixed surface through the same closed curve, a volume of given magnitude.

20. THE GENERAL PROBLEM OF BOUNDARY VALUES

An important problem closely connected with the foregoing is the question concerning the existence of solutions of partial differential equations when the values on the boundary of the region are prescribed. This problem is solved in the main by the keen methods of H. A. Schwarz, C. Neumann, and Poincaré for the differential equation of the potential. These methods, however, seem to be generally not capable of direct extension to the case where along the boundary there are prescribed either the differential coefficients or any relations between these and the values of the function. Nor can they be extended immediately to the case where the inquiry is not for potential surfaces but, say, for surfaces of least area, or surfaces of constant positive gaussian curvature, which are to pass through a prescribed twisted curve or to stretch over a given ring surface. It is my conviction that it will be possible to prove these existence theorems by means of a general principle whose nature is indicated by Dirichlet's principle. This general principle will then perhaps enable us to approach the question: *Has not every regular variation problem a solution, provided certain assumptions regarding the given boundary conditions are satisfied* (say that the functions concerned in these boundary conditions are continuous and have in sections one or more derivatives), *and provided also if need be that the notion of a solution shall be suitably extended?*[46]

---

[46]Cf. my lecture on Dirichlet's principle in the *Jahresber. d. Deutschen Math.-Vereinigung*, vol. 8 (1900), p. 184.

21. PROOF OF THE EXISTENCE OF LINEAR DIFFERENTIAL
EQUATIONS HAVING A PRESCRIBED MONODROMIC GROUP

In the theory of linear differential equations with one independent variable z, I wish to indicate an important problem, one which very likely Riemann himself may have had in mid. This problem is as follows: *To show that there always exists a linear differential equation of the Fuchsian class, with given singular points and monodromic group.* The problem requires the production of n functions of the variable z, regular throughout the complex z plane except at the given singular points; at these points the functions may become infinite of only finite order, and when z describes circuits about these points the functions shall undergo the prescribed linear substitutions. The existence of such differential equations has been shown to be probable by counting the constants, but the rigorous proof has been obtained up to this time only in the particular case where the fundamental equations of the given substitutions have roots all of absolute magnitude unity. L. Schlesinger has given this proof,[47] based upon Poincaré's theory of the Fuchsian ζ-functions. The theory of linear differential equations would evidently have a more finished appearance if the problem here sketched could be disposed of by some perfectly general method.

22. UNIFORMIZATION OF ANALYTIC RELATIONS BY MEANS
OF AUTOMORPHIC FUNCTIONS

As Poincaré was the first to prove, it is always possible to reduce any algebraic relation between two variables to uniformity by the use of automorphic functions of one variable. That is, if any algebraic equation in two variables be given, there can always be found for these variables two such single valued automorphic functions of a single variable that their substitution renders the given algebraic equation an identity. The generalization of this fundamental theorem to any analytic non-algebraic relations whatever between two variables has likewise been attempted with success by Poincaré,[48] though by a way entirely different from that which served him in the special problem first mentioned. From Poincaré's proof of the possibility of reducing to uniformity an arbitrary analytic relation between two variables, however, it does not become apparent whether the resolving functions can be determined to meet certain additional conditions. Namely, it is not shown whether the two single valued functions of the one new variable can be so chosen that, while this variable traverses the *regular* domain of those functions, the totality of all regular points of the given analytic field are actually reached and represented. On the contrary it seems to be the case, from Poincaré's investigations, that there are beside the branch points certain others, in general infinitely many other discrete exceptional points of the analytic field, that can be reached only by making the new variable approach certain limiting points of the functions. *In view of the fundamental importance of Poincaré's formulation of the question it seems to me that an elucidation and resolution of this difficulty is extremely desirable.*

In conjunction with this problem comes up the problem of reducing to uniformity an algebraic or any other analytic relation among three or more complex variables—a problem which is known to be solvable in many particular cases. Toward the

---

[47] Handbuch der Theorie der linearen Differentialgleichungen, vol. 2, part 2, No. 366.

[48] *Bull. de la Soc. Math. de France*, vol. 11 (1883).

solution of this the recent investigations of Picard on algebraic functions of two
variables are to be regarded as welcome and important preliminary studies.

23. FURTHER DEVELOPMENT OF THE METHODS
OF THE CALCULUS OF VARIATIONS

So far, I have generally mentioned problems as definite and special as possible,
in the opinion that it is just such definite and special problems that attract us the
most and from which the most lasting influence is often exerted upon science. Nevertheless, I should like to close with a general problem, namely with the indication
of a branch of mathematics repeatedly mentioned in this lecture—which, in spite
of the considerable advancement lately given it by Weierstrass, does not receive
the general appreciation which, in my opinion, is its due—I mean the calculus of
variations.[49]

The lack of interest in this is perhaps due in part to the need of reliable modern
text-books. So much the more praiseworthy is it that A. Kneser in a very recently
published work has treated the calculus of variations from the modern points of
view and with regard to the modern demand for rigor.[50]

The calculus of variations is, in the widest sense, the theory of the variation of
functions, and as such appears as a necessary extension of the differential and integral calculus. In this sense, Poincaré's investigations on the problem of three bodies,
for example, form a chapter in the calculus of variations, in so far as Poincaré derives
from known orbits by the principle of variation new orbits of similar character.

I add here a short justification of the general remarks upon the calculus of
variations made at the beginning of my lecture.

The simplest problem in the calculus of variations proper is known to consist in
finding a function y of a variable x such that the definite integral

$$J = \int_a^b F(y_x, y; x)\, dx, \quad y_x = \frac{dy}{dx}$$

assumes a minimum value as compared with the values it takes when y is replaced
by other functions of x with the same initial and final values.

The vanishing of the first variation in the usual sense

$$\delta J = 0$$

gives for the desired function y the well-known differential equation

(1)
$$\frac{dF_{y_x}}{dx} - F_y = 0,$$

$$\left[F_{y_x} = \frac{\partial F}{\partial y_x}, \quad F_y = \frac{\partial F}{\partial y} \right].$$

---

[49]Text-books: Moigno-Lindelöf, Leçons du calcul des variations, Paris, 1861, and A. Kneser,
Lehrbuch der Variations-rechnung, Braunschweig, 1900.

[50]As an indication of the contents of this work, it may here be noted that for the simplest
problems Kneser derives sufficient conditions of the extreme even for the case that one limit of
integration is variable, and employs the envelope of a family of curves satisfying the differential
equations of the problem to prove the necessity of Jacobi's conditions of the extreme. Moreover,
it should be noticed that Kneser applies Weierstrass's theory also to the inquiry for the extreme
of such quantities as are defined by differential equations.

In order to investigate more closely the necessary and sufficient criteria for the occurrence of the required minimum, we consider the integral

$$J^* = \int_a^b \{F + (y_x - p)F_p\}\, dx,$$

$$\left[F = F(p, y; x), F_p = \frac{\partial F(p, y; x)}{\partial p} \right].$$

Now we inquire how p is to be chosen as function of x, y in order that the value of this integral J* shall be independent of the path of integration, i.e., of the choice of the function y of the variable x. The integral J^* has the form

$$J^* = \int_a^b \{Ay_x - B\}\, dx,$$

where A and B do not contain y_x and the vanishing of the first variation

$$\delta J^* = 0$$

in the sense which the new question requires gives the equation

$$\frac{\partial A}{\partial x} + \frac{\partial B}{\partial y} = 0,$$

i.e., we obtain for the function p of the two variables x, y the partial differential equation of the first order

(1*) $$\frac{\partial F_p}{\partial x} + \frac{\partial(pF_p - F)}{\partial y} = 0.$$

The ordinary differential equation of the second order (1) and the partial differential equation (1*) stand in the closest relation to each other. This relation becomes immediately clear to us by the following simple transformation

$$\delta J^* = \int_a^b \{F_y \delta y + F_p \delta p + (\delta y_x - \delta p)F_y + (y_x - p)\delta F_p\}\, dx$$

$$= \int_a^b \{F_y \delta y + \delta y_x F_p + (y_x - p)\delta F_p\}\, dx$$

$$= \delta J + \int_a^b (y_x - p)\delta F_p\, dx.$$

We derive from this, namely, the following facts: If we construct any *simple* family of integral curves of the ordinary differential equation (1) of the second order and then form an ordinary differential equation of the first order

(2) $$y_x = p(x, y)$$

which also admits these integral curves as solutions, then the function $p(x, y)$ is always an integral of the partial differential equation (1*) of the first order; and conversely, if $p(x, y)$ denotes any solution of the partial differential equation (1*) of the first order, all the non-singular integrals of the ordinary differential equation (2) of the first order are at the same time integrals of the differential equation (1) of the second order, or in short if $y_x = p(x, y)$ is an integral equation of the first order of the differential equation (1) of the second order, $p(x, y)$ represents an integral of the partial differential equation (1*) and conversely; the integral curves of the ordinary differential equation of the second order are therefore, at the same time, the characteristics of the partial differential equation (1*) of the first order.

In the present case we may find the same result by means of a simple calculation; for this gives us the differential equations (1) and (1*) in question in the form

(1) $$y_{xx}F_{y_xy_x} + y_xF_{y_xy} + F_{y_xx} - F_y = 0,$$

(1*) $$(p_x + pp_y)F_{pp} + pF_{py} + F_{px} - F_y = 0,$$

where the lower indices indicate the partial derivatives with respect to x, y, p, y_x. The correctness of the affirmed relation is clear from this.

The close relation derived before and just proved between the ordinary differential equation (1) of the second order and the partial differential equation (1*) of the first order, is, as it seems to me, of fundamental significance for the calculus of variations. For, from the fact that the integral J^* is independent of the path of integration it follows that

(3) $$\int_a^b \{F(p) + (y_x - p)F_p(p)\} \, dx = \int_a^b F(\overline{y}_x) \, dx,$$

if we think of the left hand integral as taken along any path y and the right hand integral along an integral curve $\overline{y}$ of the differential equation

$$\overline{y}_x = p(x, \overline{y}).$$

With the help of equation (3) we arrive at Weierstrass's formula

(4) $$\int_a^b F(y_x) \, dx - \int_a^b F(\overline{y}_x) \, dx = \int_a^b E(y_x, p) \, dx,$$

where E designates Weierstrass's expression, depending upon y_x, p, y, x,

$$E(y_x, p) = F(y_x) - F(p) - (y_x - p)F_p(p).$$

Since, therefore, the solution depends only on finding an integral $p(x, y)$ which is single valued and continuous in a certain neighborhood of the integral curve $\overline{y}$, which we are considering, the developments just indicated lead immediately—without the introduction of the second variation, but only by the application of the polar process to the differential equation (1)—to the expression of Jacobi's condition and to the answer to the question: How far this condition of Jacobi's in conjunction with Weierstrass's condition $E > 0$ is necessary and sufficient for the occurrence of a minimum.

The developments indicated may be transferred without necessitating further calculation to the case of two or more required functions, and also to the case of a double or a multiple integral. So, for example, in the case of a double integral

$$J = \int F(z_x, z_y, z; x, y) \, d\omega, \quad \left[z_x = \frac{\partial z}{\partial x}, \quad z_y = \frac{\partial z}{\partial y}\right]$$

to be extended over a given region ω, the vanishing of the first variation (to be understood in the usual sense)

$$\delta J = 0$$

gives the well-known differential equation of the second order

(I) $$\frac{dF_z}{dx} + \frac{dF_{z_y}}{dy} - F_z = 0,$$

$$\left[F_{z_x} = \frac{\partial F}{\partial z}, F_z = \frac{\partial F}{\partial z_y}, F_z - \frac{\partial F}{\partial z}\right],$$

for the required function z of x and y.

On the other hand we consider the integral

$$J^* = \int \{F + (z_x - p)F_p + (z_y - q)F_q\}\, d\omega,$$

$$\left[F = F(p, q, z; x, y),\, F_p = \frac{\partial F(p, q, z; x, y)}{\partial p},\quad F_q = \frac{\partial F(p, q, z; x, y)}{\partial q}\right],$$

and inquire, how p and q are to be taken as functions of x, y and z in order that the value of this integral may be independent of the choice of the surface passing through the given closed twisted curve, i.e., of the choice of the function z of the variables x and y.

The integral J^* has the form

$$J^* = \int \{Az_x + Bz_y - C\}\, d\omega$$

and the vanishing of the first variation

$$\delta J^* = 0,$$

in the sense which the new formulation of the question demands, gives the equation

$$\frac{\partial A}{\partial x} + \frac{\partial B}{\partial y} + \frac{\partial C}{\partial z} = 0,$$

i.e., we find for the functions p and q of the three variables x, y and z the differential equation of the first order

$$\frac{\partial F_p}{\partial x} + \frac{\partial F_q}{\partial y} + \frac{\partial (pF_p + qF_q - F)}{\partial x} = 0.$$

If we add to this differential equation the partial differential equation

(I*) $$p_y + qp_z = q_x + pq_z,$$

resulting from the equations

$$z_x = p(x, y, z),\quad z_y = q(x, y, z),$$

the partial differential equation (I) for the function z of the two variables x and y and the simultaneous system of the two partial differential equations of the first order (I*) for the two functions p and q of the three variables x, y, and z stand toward one another in a relation exactly analogous to that in which the differential equations (1) and (1*) stood in the case of the simple integral.

It follows from the fact that the integral J^* is independent of the choice of the surface of integration z that

$$\int \{F(p, q) + (z_x - p)F_p(p, q) + (z_y - q)F_q(p, q)\}\, d\omega$$

$$= \int F(\bar{z}_x, \bar{y})\, d\omega,$$

if we think of the right hand integral as taken over an integral surface $\bar{z}$ of the partial differential equations

$$\bar{z}_x = p(x, y, \bar{z}),\ \bar{z}_y = q(x, y, \bar{z});$$

and with the help of this formula we arrive at once at the formula

(IV) $$\int F(z_x, z_y)d\omega - \int F(\bar{z}_x, \bar{z}_y)\, d\omega = \int E(z_x, z_y, p, q)\, d\omega,$$

$$[E(z_x, z_y, p, q) = F(z_x, z_y) - F(p, q) - (z_x - p)F_p(p, q)$$
$$-(z_y - q)F_q(p, q)],$$

which plays the same rôle for the variation of double integrals as the previously given formula (4) for simple integrals. With the help of this formula we can now answer the question how far Jacobi's condition in conjunction with Weierstrass's condition $E > 0$ is necessary and sufficient for the occurrence of a minimum.

Connected with these developments is the modified form in which A. Kneser,[51] beginning from other points of view, has presented Weierstrass's theory. While Weierstrass employed to derive sufficient conditions for the extreme tehos integral curves of equation (1) which pass through a fixed point, Kneser on the other hand makes use of any simple family of such curves and constructs for every such family a solution, characteristic for that family, of that partial differential equation which is to be considered as a generalization of the Jacobi-Hamilton equation.

The problems mentioned are merely samples of problems, yet they will suffice to show how rich, how manifold and how extensive the mathematical science of to-day is, and the question is urged upon us whether mathematics is doomed to the fate of those other sciences that have split up into separate branches, whose representatives scarcely understand one another and whose connection becomes ever more loose. I do not believe this nor wish it. Mathematical science is in my opinion an indivisible whole, an organism whose vitality is conditioned upon the connection of its parts. For with all the variety of mathematical knowledge, we are still clearly conscious of the similarity of the logical devices, the *relationship* of the *ideas* in mathematics as a whole and the numerous analogies in its different departments. We also notice that, the farther a mathematical theory is developed, the more harmoniously and uniformly does its construction proceed, and unsuspected relations are disclosed between hitherto separate branches of the science. So it happens that, with the extension of mathematics, its organic character is not lost but only manifests itself the more clearly.

But, we ask, with the extension of mathematical knowledge will it not finally become impossible for the single investigator to embrace all departments of this knowledge? In answer let me point out how thoroughly it is ingrained in mathematical science that every real advance goes hand in hand with the invention of sharper tools and simpler methods which at the same time assist in understanding earlier theories and cast aside older more complicated developments. It is therefore possible for the individual investigator, when he makes these sharper tools and simpler methods his own, to find his way more easily in the various branches of mathematics than is possible in any other science.

The organic unity of mathematics is inherent in the nature of this science, for mathematics is the foundation of all exact knowledge of natural phenomena. That it may completely fulfil this high mission, may the new century bring it gifted masters and many zealous and enthusiastic disciples.

---

[51] Cf. his above-mentioned textbook, §§14, 15, 19 and 20.